HEAT TRANSFER VIRTUAL LAB FOR STUDENTS AND ENGINEERS

HEAT TRANSFER VIRTUAL LAB FOR STUDENTS AND ENGINEERS

THEORY AND GUIDE FOR SETTING UP

ELLA FRIDMAN AND
HARSHAD S. MAHAJAN

MOMENTUM PRESS

MOMENTUM PRESS, LLC, NEW YORK

Heat Transfer Virtual Lab for Students and Engineers: Theory and Guide for Setting Up
Copyright © Momentum Press®, LLC, 2014.

First published by Momentum Press®, LLC
222 East 46th Street, New York, NY 10017
www.momentumpress.net

ISBN-13: 978-1-60650-548-9 (print)
ISBN-13: 978-1-60650-549-6 (e-book)

Momentum Press Thermal Science and Energy Engineering Collection

DOI: 10.5643/9781606505496

Cover design by Jonathan Pennell
Interior design by Exeter Premedia Services Private Ltd., Chennai, India

10 9 8 7 6 5 4 3 2 1

Printed in the United States of America

ABSTRACT

Laboratory experiments are a vital part of engineering education, which historically were considered impractical for distance learning. In view of this, the proposed book presents a guide for the practical employment of a heat transfer virtual lab for students and engineers. The main objective of our virtual lab is to design and implement a real-time, robust, and scalable software system that provides easy access to lab equipment anytime and anywhere over the Internet. We have combined Internet capabilities with traditional laboratory exercises to create an efficient environment to carry out interactive, online lab experiments. Thus, the virtual lab can be used from a remote location as a part of a distance learning strategy. Our system is based on client-server architecture. The client is a general purpose java-enabled web-browser (e.g. Internet Explorer, Firefox, Chrome, Opera, etc.) which communicates with the server and the experimental setup. The client can communicate with the server and the experimental setup in two ways: either by means of a web browser, which runs a dedicated CGI (Common Gateway Interface) script in the server, or using the LabVIEW Player, which can be downloaded and installed for free. In both cases, the client will be capable of executing VIs (Virtual Instruments) specifically developed for the experiment in question, providing the user with great ability to control the remote instrument and to receive and present the desired experimental data. Examples of this system for several particular experiments are described in detail in the book.

KEY WORDS

armfield, distance learning, engineering education, heat exchanger, heat transfer, heat transfer laboratory experiments, HT-30xc CGI (Common Gateway Interface) script, LabVIEW Player, online lab experiments, remote instruments, virtual laboratory, VIs (Virtual Instruments)

CONTENTS

LIST OF FIGURES

ACKNOWLEDGMENTS

I would like to express my sincere thanks to my guide Dr. Ella Fridman for her continuous support and encouragement throughout the duration of the project. Her timely suggestions and patience in difficult phases of my project were instrumental in successful completion of the project. Special thanks to Dr. Clark Colton for giving me access to the online experiments at Massachusetts Institute of Technology. The actual online-experiment experience helped me understand my project better.

Successful completion of my project would not have been possible without the initial help of my senior, Viji Venkatachalan. Her initial insight into the project, as well as helping me grasp the "LabVIEW web publishing tool," saved me a good deal of project time.

Sincere thanks to Mr. Allen Rioux for providing me with the required information for FTP access on the web server. Thanks to Al for extending his help in fixing the mechanical problems with the equipment. Last but not least, I am also grateful to all the students who used this project to perform their lab experiments and provided us with their invaluable feedback that helped improve the project.

CHAPTER 1

INTRODUCTION

The objective of this project is to design a user-friendly and efficient system for interactive, online operation of remote education laboratory equipment and experiments utilizing the Internet. This chapter introduces the concept and short history of Virtual Laboratory and discusses its needs and advantages. As a part of the requirements analysis for the Virtual Lab project, we touch upon topics such as the overall system architecture and model hierarchy, role design, and web user interface design. This chapter lays the foundation for the rest of the book.

1.1 HISTORY OF DISTANCE LEARNING AND CONCEPT OF VIRTUAL LAB

During the last decade, we have witnessed rapid developments of computer networks and Internet technologies along with dramatic improvements in the processing power of personal computers. These developments make interactive distance education a reality. By designing and deploying distributed and collaborative applications running on computers disseminated over the Internet, distance educators can reach remote learners, overcoming the time and distance constraints.

Besides the necessary theoretical base provided by lectures and written materials, hands-on experience provided by physical laboratories is a vital part of engineering education. It helps engineering students become effective professionals. It not only provides students the knowledge of the physical equipment but also adds the important dimension of group work and collaboration. However, laboratories are expensive to setup, to maintain, and to provide long hours of daily staffing. Due to budget limitations, many universities and colleges can provide only limited access to such physical equipment. Therefore, it is imperative to enable remote access to a physical laboratory, as part of either an onsite or distance-learning course.

Students like it better if given a chance to collaborate in small learning groups. They are more motivated if they are in frequent contact with the instructor. Therefore, the incorporation and reinforcement of collaboration and interaction to support real-time video and audio communication are becoming important features of remote laboratories. The objective of this project is to design a user-friendly and efficient system for the interactive, online operation of remote education laboratory equipment and experiments, utilizing the Internet. Our goal is to introduce a remote lab design that is simple, scalable, and flexible enough to allow users, who may not be computer experts, to use the system to conduct virtual experiments. The particular focus of this work is the design of a real-time, robust, and scalable software system for use in thermodynamics courses to provide students with hands-on experience with a heat transfer experiment for them to compare measured characteristics with theoretical predictions and reflect on discrepancies, limitations, and design constraints. The system must run 24×7 to allow students to access it round the clock.

The heat exchanger unit used was a general-purpose service unit, designed by Armfield Ltd, which supplied facilities and infrastructure and was used in conjunction with a range of small-scale accessory equipment for carrying out specific experiments involving heat exchangers. The service units are operated and controlled via LabVIEW software on a computer that, in turn, functions as a server with LabVIEW software and that enables monitoring and control of the experiment on the Internet. For remote operation of the experiment, it requires a browser and plug-in that supports the Java 2 Runtime Environment (or the preceding) and the LabVIEW 7.0 (or the preceding) Runtime Environment.

1.2   WHAT IS VIRTUAL LAB?

The main objective of a Virtual Lab is to design and implement a real-time, robust, and scalable software system around laboratory equipment that provides a "learner" an easy access to the lab equipment anytime and anywhere over the Internet.

The Internet offers interesting possibilities for disseminating educational material to students, both locally and as part of remote education. Laboratory experiments are a vital part of engineering education, which have so far been considered impractical for distance learning. However, recent advances in Internet technologies and computer-controlled instrumentation presently permit Internet-based techniques to be utilized for setting up remote laboratory access. Also, the use of Internet and studio classrooms is an emerging trend for promoting "individual discovery" as a strategy for enhancing engineering education.

Here, we describe how these techniques can be combined with traditional laboratory exercises to create an efficient environment for interactive online operation of lab experiments over the Internet, to be used either in a studio setting or from a remote location as part of a distance-learning strategy. Our system is based on client-server architecture. The client is a general purpose java-enabled web browser, for example, Internet Explorer, Firefox, Chrome, Opera, and so forth. Web browsers communicate with the server and the experimental setup. In some cases custom desktop software can be used as a client.

Previous versions of remote lab were based on a transmission control protocol/Internet protocol (TCP/IP) solution, which used Java applet technology on the client (i.e., student) side. This was achieved by virtue of a Java virtual machine (JVM) in the web browser that could download and execute Java applet code. The client would see a pop-up window that provided interaction and communication directly with the server. However, unsigned applets make it tough for the client to store and present the measurement data, and to transfer them to other applications (except by "cut-and-paste") because of Java's security structure. An intermittent problem with Java applet is that the functionality of an applet may vary between different browsers.

The LabVIEW software from National Instruments offers an interesting solution for measurement and control system. It also provides the desired Internet access to the lab, out of the box. It has the following interesting features:

1. Graphical programming
2. Simplicity in design
3. Acquire and save the measurements and readings for further analysis in various file formats including xls, csv, txt, and so forth
4. Stand-alone instrument control through vendor-specific or generic plugins
5. Automated tests and validation system

In this solution, the lab-side server runs a "full version" of LabVIEW, which incorporates Internet communication capabilities and functionalities to access/control instruments and to acquire/output data. The client can communicate with the server and the experimental setup in two ways: either by means of a web browser, which runs a dedicated Common Gateway Interface script in the server, or by using the LabVIEW Player, which can be downloaded and installed for free. In both cases, the client will be capable of executing virtual instruments (VIs) specifically developed for the experiment in question, providing the client with great ability to control the remote instrument and to receive and present the desired experimental data. The other solution seeks to exploit the additional functionally of the recent browsers, enabling the server system to respond in many different formats, such as JavaScript, HTML, or eXtensible Markup Language (XML), which gives the client great flexibility in storing, processing, and presenting the data received. This is achieved by creating web solutions based on either the information server application information interface (ISAPI) server extensions, or on a component object model with extensions (COM+) solution at the lab side.

1.3   ANALYSIS OF PROJECT REQUIREMENTS

The project intends to increase its student base through online education aimed at fulfilling the needs of remote students. Remote students need time flexibility, instant guidance, and feedback. This project intends to design a system around virtual labs to achieve exactly that in a perfect pedagogical approach. The project places great emphasis on laboratories that account for approximately 40 percent of program content. The distance-learning program must continue to offer the same quality of interaction with the faculty and the laboratory that it now offers its onsite students. Remote laboratories have been successfully used in electrical engineering education to interact with spectroscopy, measurements, and control systems laboratories. The same is to be achieved for mechanical engineering students.

This book describes the pilot version of a remote interactive laboratory that is used for thermodynamics laboratories by students from remote sites. In a remote delivery scenario, it is important that the delivery mechanism, laboratory course content, and instructional design be tailored to

1. Model an active remote-learning environment that engages the student in achieving learning outcomes
2. Model a collaborative environment for group interactions
3. Design appropriate roles for supporting the collaborative environment
4. Provide unambiguous feedback and instant guidance to the students
5. Match the characteristics of the media (delivery medium) to specific learning outcomes and processes

A discussion of the salient pedagogical features of the onsite program is provided as this is important to understand the requirements and implementation issues faced by the online program. The Virtual Lab employs three types of interactions to ensure effective learning:

1. Lectures by expert instructors
2. Hands-on laboratories
3. Group interaction with peers

Case studies, projects and group work correlate with the three well-known pedagogical approaches, namely, objectivist, constructivist, and group interaction, respectively. The objectivist approach emphasizes that students learn by explicitly being informed or taught by subject experts. The constructivist approach is based on learners learning by performing authentic activities and constructing knowledge in authentic learning environments.

The group interaction approach is based on groups of learners engaging in collaborative problem solving that increases student engagement with the subject matter resulting in better learning. The hands-on laboratories build practical internetworking abilities and skills in students and correspond to a constructive, collaborative, situated, learner-centric environment. Situated learning has been used in technology-based courses to present academic knowledge in a practical context to teach students problem-solving skills and is employed in the Virtual Lab to transform novice students into experts in the context of the industry in which they will ultimately work.

In the labs, the students gain a broad range of hands-on experience and knowledge to understand the practical conditions under which to apply specific internetworking principles, theories, and techniques. The laboratory employs state-of-the-art networking equipment, simulators, and other hardware in a learner-centric environment that engages students in collaborative activities. Students learn to apply theoretical knowledge to practical networking issues, hands-on configuration of equipment, and strategies and techniques for troubleshooting networks. Most engineering activities in a modern enterprise are conducted in a collaborative setting with a lot of interaction among team members. This makes it imperative that the project models and implements a collaborative environment onsite to facilitate the acquisition of problem solving, reasoning, and management skills required by the companies today. Student interaction is encouraged by suitably designing laboratory activities such that students carry them out in groups of two to three.

1.4 LEARNING THEORY AND ITS INFLUENCE ON ROLE DESIGN

In addition to a rich repertoire of learning resources and aids, e-learning includes tailoring learning modules to address how students engage in learning, fostering effective e-learning strategies and instructional design that incorporate the latest techniques in pedagogical research to support learning at a pace that is comfortable to the student. These objectives can be met with either a self-paced environment in which the student learns at a rate comfortable to the individual or a directed environment in which the student has to follow a particular sequence of instructions.

The learning environment can also be classified as synchronous requiring the simultaneous participation of students in the class or asynchronous in which a student may participate at a time convenient to them. Also, the characteristics of a media used in communication can be assessed using media synchronicity theory and include characteristics such as a medium's capacity to provide feedback, symbol variety, instruction of multiple students, tuning message content, extent to which message can be reprocessed, and unambiguousness.

One of the challenges facing the online laboratory is how best to mimic the onsite face-to-face interaction between students and faculty, which is critical to learning technology-intensive courses. Most onsite students benefit from face-to-face interaction with instructors, provided the faculty to student ratio is at reasonable levels. The same interaction can be achieved in an online program using a well-designed facilitation approach. Good e-learning begins with effective, real-time, reliable, and secure student interaction with the e-learning system. All other steps in the e-learning process rely on this crucial student interaction phase. The most important measure that a student will use for repeat interaction with an e-learning system of a university is the ease in using the system.

From the universities' viewpoint, the ease in using an e-learning system is a function of system design and is determined by several factors such as its accessibility, usability, reliability of system, help available, responsiveness of the system, and appropriateness of system response to student input and support for many simultaneous users. The communication channel characteristics, protocols, and technology must be designed for real-time applications. The metrics by which an e-learning resource may be evaluated include the following:

1. Expert curriculum
2. Ease of use

3. Involve continuous assessment
4. Allow real-time feedback to track student performance
5. Employ multimedia simulations, laboratories, and user interaction to create a dynamic, engaging environment for learning
6. Enhance problem-solving techniques on an individual or group basis

Most courses require the students to interact with the devices in the laboratory. For this purpose, onsite students access and configure the devices in the laboratory using a command line interface or a graphical user interface (GUI). A key issue with the remote delivery of the Virtual Lab content is to convert the onsite student interaction with the devices in the laboratory into online real-time interaction with the devices.

1.5 SYSTEM ARCHITECTURE

It is prototype system that is intended to support experiments in the area of electronics, but its structure poses no limitations on other types of experiments also, such as in physics, mechanical engineering, or some similar field of engineering or science.

It is quite a general structure that can be implemented using various software and hardware in different ways. The main goal and intention of such configuration is to fulfill a number of requirements.

- Universal configuration suitable and easily adopted to various kinds of experiments
- Structured configuration that can be implemented using various hardware and software
- Easily scalable configuration that can be extended to meet the following requirements:
  1. Increase in number of users
  2. Introduction of new experiments
  3. Increased number of existing experiment runs
- Consists of hardware/software that are not too specific and intended more or less for general use
- Control of users competing requests for the same experiment or the same experiment equipment

Clear distinction and defined communication between classes on different levels aids to minimum dependence between classes and also enables reusability and independent changes between classes, as long as interfaces

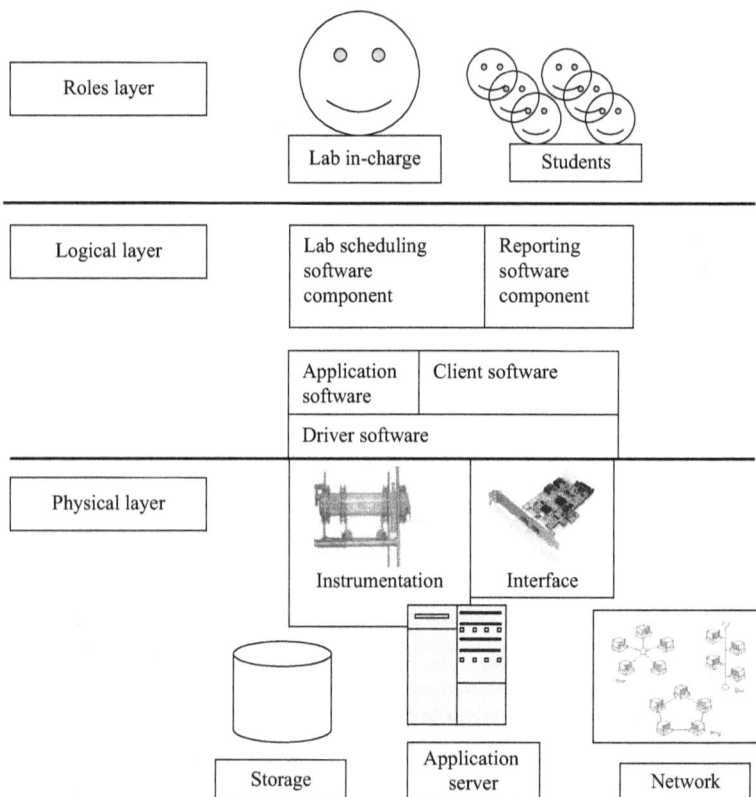

Figure 1.1. Key elements of a Virtual Lab.

between classes remain the same. Remote access to experiment and experimental equipment has various limitations and demands that can be met up to a certain extent. The main idea and intention of this system is to make and provide the basic functionality of laboratory conditions and requirements usual for practical and experimental work, online, from the side of a remote student that uses a PC and Internet browser as the main and only tool.

Two basic approaches are possible. One is the intention to make very similar conditions and user interface to instruments that are in the real lab, to make available all controls, switches, and buttons existing in real equipment, also available for online student. Such an approach is a very demanding one, and it is essentially impossible to create identical conditions for a remote user and student in the real lab working on real equipment.

The other approach is a more flexible one, that is, the intention is not to create as similar conditions as possible, but to make some kind of adapted remote laboratory system that will provide the functionality based on the logical interconnection with theoretical material, serving as support for easier understanding and also for the introduction and illustration of basic notions characteristic for measurements. The presentation and user interface are accommodated to different conditions imposed by remote access. Only the most important controls, measured values, and results are shown, and the user interface can be quite different comparing to laboratory instruments, but logically quite functional and sufficient for complete understanding. In this book, the first approach is used, as it is primarily for the illustration of main characteristics, functionality, and capabilities of implemented prototype of remote laboratory system.

1.6   MODEL HIERARCHY

In Figure 10.1 Roles Layer can be further divided into 3 tiers. Tier 1 consists of "Faculty and Administration" (including the director), Tier 2 consists of teaching/lab assistants, and Tier 3 consists of students. Tier 1 administration handles finances, enrollment, registration, and other functions associated with disseminating program information (students get automatically enrolled for the lab class when they register for the course associated with it). The faculty is the sole course content provider in charge of designing an expert curriculum. It also administers tests, examines and assesses students, and provides feedback on student competencies, thus meeting the e-learning resource metrics of expert curriculum. It is also the responsibility of Tier 1 personnel to maintain the integrity of the educational process. The teaching/lab assistants (Tier 2) maintain and update lab notes for each course. In addition, they test and configure the devices in the Internet-working laboratory for proper use and create and maintain user account information based on information from the administration. In general, the teaching/lab assistants, guided by the faculty, maintain a dynamic, engaging mechanical laboratory that is easy to use and meet e-learning resource metrics. The Tier 2 support maintains and upgrades network services on servers and workstations at the remote site.

1.7   WEB USER INTERFACE

Web user interface is on the highest level and serves for communication with remote users—students. Web interface for Virtual Lab consists of two parts. The first part is the theoretical introduction and explanation for the experiment that will be performed. It also includes precise instructions on how to perform the experiment, the meaning of all values that should be entered or selected, and values that are obtained as a result of the experiment. Besides textual information about the experiment, it supports a graphical presentation of measured data in the experiment, which is important for the full understanding of the performed experiment. With graphical presentation, it is very easy to see that presented data are from the real world, obtained on real equipment with all the influences that cannot be avoided in real experiment. Together with the basic shapes of obtained lines that characterize the observed dependencies, the unavoidable noise is also present. A web interface should also contain means for control of user access to the system.

1.7.1 EXPERIMENT

The logic is used for the following:

- Collecting the user input data from the corresponding web page
- Checking the data consistency
- Preparing and sending data to programmable devices for equipment management
- Full control and management of the experiment process using the programmable devices
- Accepting the data measured in experiment from programmable devices
- Data processing and calculation of indirectly measured quantities
- Preparing and sending data to a web server

1.8 QUESTIONS

1. **What is the main objective of a "Virtual Lab"?**
   The main objective of a Virtual Lab is to design and implement a real-time, robust, and scalable software system around laboratory equipment that provides a "learner" easy access to the lab equipment anytime and anywhere over the Internet.
2. **List the technologies that enable "Virtual Lab".**
   a. Internet/web technologies and computer-controlled instrumentation
   b. Java, Java Virtual Machine (JVM), Java Runtime Environment JRE
   c. LabVIEW

3. **What is the problem with using applets for designing a Virtual Lab system?**

   When a user visits the experiment website and if the page has applet to run, the user is presented with a confirmation dialogue. The applet code is run if and only if the user permits. Sun Systems designed the security model of applet in such a way that not even a novice user is duped into running a malicious code on the computer, without his own notice.

   There are two types of applets:

   1. Sandbox applets or unsigned applets
   2. Privileged applets or signed applets

   Sandbox applets are run in a security sandbox that allows only a set of safe operations. Sandbox applets are not signed. Their limitations are (but not limited to) as follows:

   - They cannot access client resources such as the local filesystem, executable files, system clipboard, and printers. Thus this makes it hard for the user to save the experiment data.
   - They cannot connect to or retrieve resources from any third-party server (any server other than the server it originated from). This puts limitations on the system design.

   In order to allow the applet to save the data to the hard disk, one must make it a privileged applet by getting the applets signed. Privileged applets can run outside the security sandbox and have extensive capabilities to access the client. Privileged applets are signed by certification authorities digitally at a cost.

4. **What is JVM?**

   JVM is, as the name suggests, a virtual machine that is capable of executing "Java bytecode." It is a software that Sun Systems has designed. Sun Systems has separate "JVMs" for individual platforms. JVM in essence acts as a translator that translates Java bytecode instructions to platform-specific instructions (Figure 1.2). What is the advantage of all this? The answer is "platform independence." It is a huge time saver for application developers! It facilitates the application, designed for one platform, to be ported to another platform with more ease. It does not need to be completely re-written. One interface designed for accessing the lab can now be used from desktops/laptops (Intel X86 JVM) as well as from mobiles or PDAs (ARM JVM).

5. **How does National Instrument's LabVIEW software help in the virtualization of a lab?**

   NI LabVIEW provides a matured solution for the virtualization of a lab. This software has existed for more than 20 years now in the

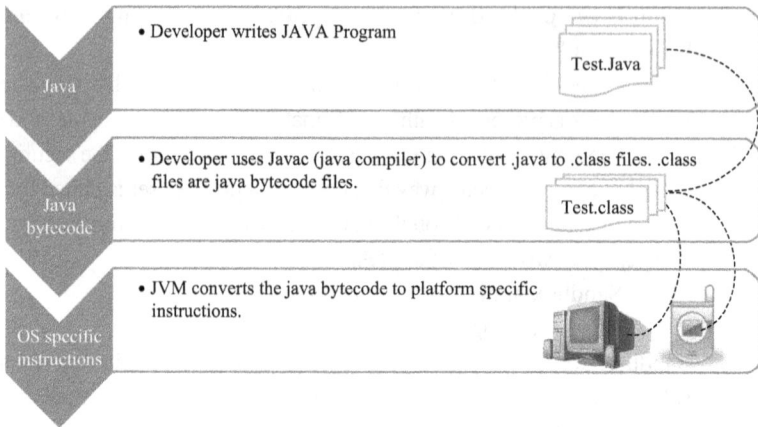

| | • Developer writes JAVA Program | Test.Java |
| Java | | |

| | • Developer uses Javac (java compiler) to convert .java to .class files. .class files are java bytecode files. | Test.class |
| Java bytecode | | |

| | • JVM converts the java bytecode to platform specific instructions. | |
| OS specific instructions | | |

Figure 1.2. How JVM functions.

scientists' and engineers' community. Instrumentation is a piece of cake with the help of LabVIEW. The graphical programming feature takes the complexity out of the design phase, thus facilitating rapid prototyping and validation. A rich plethora of plugins let the VI designer achieve anything and everything possible in the world of instrumentation.

The LabVIEW learning curve for a novice user is not steep either, thanks to the graphical programming feature, budding community of users, and exceptionally prompt support from NI. One can be quickly up to speed in a matter of days.

The feature that really enables remote access is "publishing a VI" over the Internet. It literally takes nothing but a "button click." Literally! Once a VI is published, it runs a server on a given port. It shows the address to which the clients can connect, for example, http://<your public ip>:5080. The users can connect to this address using any browser, just like we browse any other website. Once the VI is loaded completely onto the user's browser, the user can control the VI to his wish.

OS SUPPORT

LabVIEW supports the various operating systems: as listed in 01: Operating Systems supported by LabVIEW

Table 1.1. Operating Systems supported by LabVIEW

Windows	Mac OS X	Linux	Sun Solaris
Windows NT	"Tiger"	Red Hat Enterprise Linux	SPARC 32-Bit
Windows 2000	10.5 "Leopard"	Scientific Linux	
Windows XP (32-bit)	10.6 "Snow Leopard"	Open SUSE Linux	
Windows Vista (32-bit)	10.7 "Lion"		
Windows Vista (64-bit)	10.8 "Mountain Lion"		
Windows 7 (32-bit)			
Windows 7 (64-bit)			
Windows 8 (32-bit)			
Windows 8 (64-bit)			
Windows RT			
Windows Server 2003 R2 (32-Bit)			
Windows Server 2008 R2 (64-Bit)			

CHAPTER 2

LABVIEW BASICS

If the reader is already familiar with LabVIEW, this chapter can be skipped. This chapter covers the following aspects of LabVIEW keeping the scope of this book in mind:

1. Block Diagram
2. Front Panel
3. G-Language
4. LabVIEW Palettes
5. Hands on LabVIEW for designing simple calculator

2.1 LABVIEW INTRODUCTION

LabVIEW is a very nice tool for scientists, engineers, and academies to perform rapid prototyping of virtual instruments. LabVIEW programs are called as VIs, that is, virtual instruments. The LabVIEW basics can be itemized as follows:

1. Graphical language (G-language) programs called virtual instruments (VIs)
2. Each VI has two main components:
   a. Front Panel: User interface of the VI.
   b. Block Diagram: The code that contains graphical representations of functions to control the Front Panel objects. It may contain sub-VIs.
3. The Front Panel contains the following:
   a. Controls: Inputs to the program (numerals, strings, etc.)
   b. Indicators: Outputs of the program (numerals, strings, plots, etc.)
4. A Block Diagram consists of the following:

a. Icons: These are means of representing operations and sub-Vis. Icons have terminals defining inputs and outputs to the operations. The sub-Vis, representations of controls and indicators in the Front Panel, are "wired" to each other or to constants.

b. Wires: Means of connecting operations and VIs to other operations, VIs, inputs and outputs, controls, and indicators.

5. The palettes are as follows:
a. Tools
b. Controls
c. Functions

2.2 G-LANGUAGE

To engineers, scientists, researchers, and all of us, it is more intuitive to think graphically. LabVIEW is equipped with the G-language. "G," in G-language, represents "Graphical." As the name suggests, it allows the user to program, "graphically". A user can drag and drop a "for loop structure icon" instead of typing for loop syntax and thus "drawing a program in a block-diagram" instead of "typing it."

G is a complete programming language. LabVIEW comes with a full suite of compiler, linker, and debugger for G. The Block Diagram programmed using G is directly compiled into the machine code, just like any other language.

The interactive debugging tool, as shown in Figure 2.1, is the best friend of LabVIEW developer. Single stepping, probing wires, and execution highlighting are few features to name. "Execution highlighting" illuminates the wire, controls, and shows the data as it passes through the Block Diagram. This helps in understanding the order of execution. When the Block Diagram has any syntactical problems, the broken Run arrow

immediately indicates that the VI needs to be fixed. Various data types are presented with different colors, as mentioned in Table 2.1 and different shapes of wires.

It is easy to embed a C code in LabVIEW. All one needs to do is build a DLL from the C code. LabVIEW provides the Call Library Function node, which can be used to link to such a DLL and call the function.

Figure 2.1. Debugging tools in LabVIEW.

Table 2.1. Color code representation of data
types in LabVIEW

Color	Data Type
Blue	Integer
Orange	Float
Purple	Char
Green	Bool

Unfortunately embedding a Java code is not as easy as embedding C. One needs to implement a Java Native Interface (JNI). The functions calling into the JNI internally can be wrapped in a wrapper DLL. This DLL can provide an API to the external world. Wrapper DLLs can be created by making use of C IDE like LabWindows.

Let us look at the java code snippet that calculates Σ5 and how the same program can be drawn in G-language. "Drawing a program on a canvas" can be a new concept. It is very similar to drawing any Block Diagram. Each programming structure is a block, with input and output. These blocks are connected to each other using "wires." Java code and G Block Diagram are shown side-by-side in Table 2.2.

2.3 FRONT PANEL

The Front Panel is the user interface of the VI. Figure 2.2 shows a simple Front Panel of a calculator. We will design the same in Section 2.6.1.

The Front Panel is a kind of "front desk" of the company. It can be designed by bringing together various input and output controls from the palettes. The input controls include text boxes, numeric incrementers, toggle buttons, sliders, and so forth. The output controls can be graphs, LEDs, labels, and so forth. Other controls include labels, images, borders, and so forth. For every control, we pull onto the Front Panel; we get its background counterpart in the Block Diagram window. A Front Panel can be made as sophisticated as the user wants, depending upon the application. LabVIEW provides a plethora of controls, such as various types of buttons, LED indicators, dials, waveform plotters, and so forth. The Front Panel shown in the image is of a simple calculator application. It shows two numerical inputs and four numerical outputs. Each of these six controls has its counterpart in the Block Diagram window shown in the following section.

Table 2.2. Comparison of java code with LabVIEW G-language

Java	G-language

Summation.java
package com.momentumpress.labview.ch7;
publicclass Summation {
publicstaticvoid main(String[] args){
    int N = 5;
    int sum = 0;
    //For loop java syntax
    for(int i=1; i<=N; i++){
        sum = sum + i;
    }
}
}
System.*out*.println("Summation of "+ N + " = "+sum);
	Summation.vi
	
	N (Input)

Output: **Summation of 5 = 15**	**Output:**

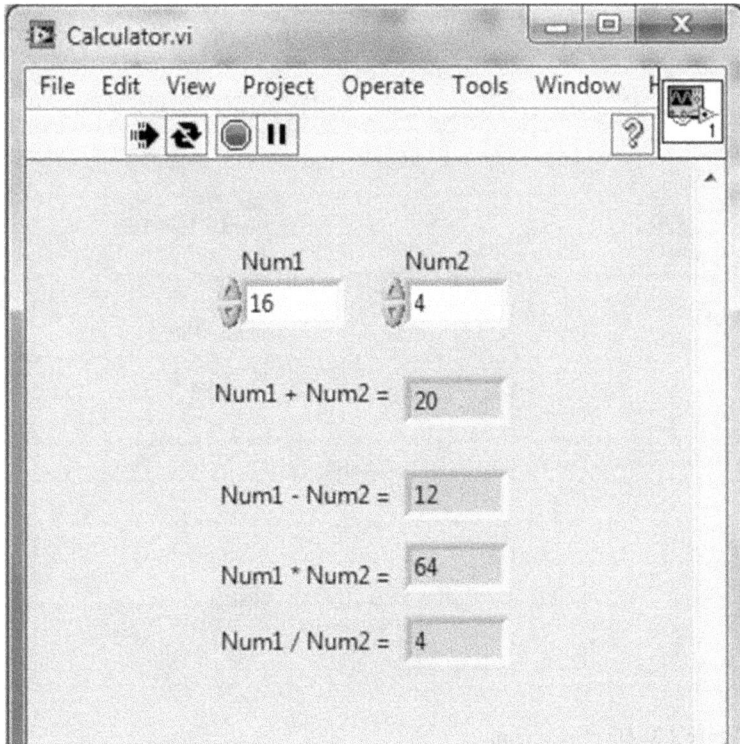

Figure 2.2. Front Panel.

Depending upon the application, Front Panels can quickly get pretty sophisticated. The Front Panel of the virtual lab experiment of Chapter 4 is a good example of a sophisticated Front Panel. It makes use of many controls, indicators, charts, and programming structures. It can be seen in Figure 4.5.

2.4 BLOCK DIAGRAM

A Block Diagram is the back office where all the processing really happens. This is where all the components, controls, and functions are wired together. The Figure 2.3 shows the Block Diagram of the Front Panel of a calculator, shown in the Figure 2.2. Num1 and Num2 are the numerical inputs, whereas the rest of the four controls are numerical outputs. The orange wires and controls represent their data type as integer. The triangles represent the numerical operations; we perform on the numerical inputs. It is fairly simple to comprehend, mainly due to its graphical nature.

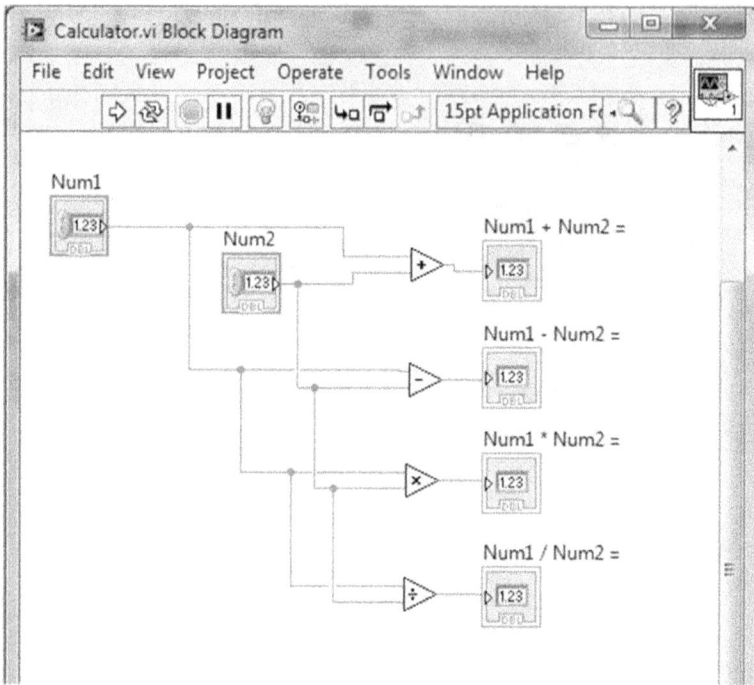

Figure 2.3. Block diagram.

2.5 LABVIEW PALETTES

2.5.1 TOOLS PALETTE

Figure 2.4 shows the Tools Palette. Select the menu option View > Tools Palette to see the Tools Palette. Hover the cursor on the small icons to see the name of each tool. In the palette, there are the following: Operating tool, Positioning tool, Labeling tool, Wiring tool, Edit Text tool, Scroll Window tool, Breakpoint tool, Probe tool, and Operate Value tool. When a tool icon is clicked, the cursor takes the shape of the icon when it is on the Front Panel or the Block Diagram, thus giving visual guidance of the tool to the user. Alternatively, if the Automatic Tool Selection is on, the mouse pointer automatically turns itself into the appropriate tool when brought closer to the relevant objects. For instance, the mouse pointer turns into a Connect Wire ⟦⟧ tool when brought closer to the terminals on the Block Diagram window, or it turns into Operate Value ⟦⟧ tool when brought closer to a button on the Front Panel window.

Figure 2.4. Tools Palette.

2.5.2 CONTROLS PALETTE

Select the menu option "View > Controls Palette" or right-click the Front Panel to see the Controls Palette (Figure 2.5). This palette has subpalettes as shown in Table 2.3 (as of LabVIEW 2013).

Notice that this palette is highlighted and operational when the Front Panel is active. Hover the cursor on the icons to see the pop-up names. Expand the subpalettes to see the available controls and indicators under that category. Show/Hide Subpalettes allows the user to pick and choose the visible subpalettes.

2.5.3 FUNCTIONS PALETTE

Select the menu "View > Functions Palette" or right-click the Block Diagram workspace to display the Functions palette (Figure 2.6). If you right-click the Block Diagram window, the Function palette appears in temporary mode, in other words, it disappears immediately when the Block Diagram loses focus. To make the Functions palette stay on, click the thumbtack in the upper left corner of the palette to pin the palette so it is no longer temporary.

Subpalettes in the Functions palette are described briefly in Table 2.4.

2.6 PROGRAMMING WITH LABVIEW

LabVIEW programming is a relatively easy task. One does not have to learn clumsy syntax. It is advisable that the user understands the basics of LabVIEW explained in the previous sections. Knowledge of LabVIEW

Table 2.3. Controls sub-palettes

Subpalettes	Description
Modern	You will find most of your Front Panel needs catered by this subpalette.
Silver	The silver controls are adaptive controls that change their look and feel depending on the platform the VI is run on.
System	Collection of controls and indicators to use in creating the container UI such as horizontal tabs and vertical tabs. It also contains decorative controls.
Classic	Collection of controls and indicators to create Vis for low-color monitor settings.
Express	It is a subset of frequently used controls from the Modern subpalette. The aim of this palette is to allow the frequently used controls to be located quickly.
.NET & ActiveX	Collection of controls and indicators to manipulate common .NET or ActiveX controls.
User Controls	This palette is initially empty. The user can custom-create a control and add it to this palette.
Control Design and Simulation	Collection of controls and indicators to construct plant and control models using transfer function, state-space, or zero-pole-gain. This category also has tools to analyze system performance with tools such as step response, pole-zero maps, and Bode plots.
Signal processing	Collection of tools for wavelet and filter-bank design for short-duration signal characterization, noise reduction, time-series analysis, and time-frequency analysis.
Add-ons	The user can purchase specific toolkits from National Instruments or third-party vendors. Such toolkits appear in this category.

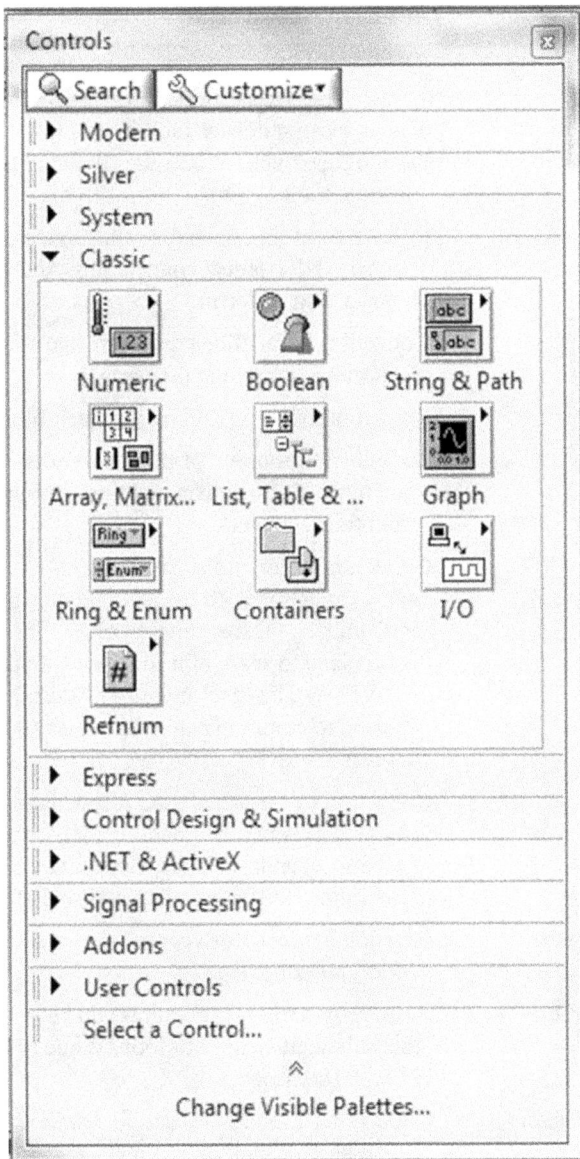

Figure 2.5. Controls palette.

Table 2.4. Subpalettes in the functions palette

Subpalettes	Description
Add-ons	The user can purchase toolkits and install it as plugins. Such toolkits are shown in this subpalette.
Connectivity	Collection of functions for web services, source control, .NET object connections, ActiveX objects, and so forth.
Data communication	Collection of functions used for Data Communication over the ports.
Express	Collection of common measurement functions.
Favorites	Convenient grouping of frequently used functions. The user can add/remove functions from this subpalette.
Instrument I/O	Collection of functions used for I/O, communication with the external physical instruments. NI Instrument Driver Finder is a good place to search for the drivers to use with LabVIEW. The Instrument I/O Assistant can be used to communicate with message-based devices.
Mathematics	Collection of functions used for mathematical analysis. It contains functions for Curve Fitting, Probability and Statistics, Linear Algebra, Integration, Differentiation, and so forth.
Measurement	Installed hardware drivers are shown as functions under this subpalette.
Programming	Basic programming structures can be found under this subpalette, e.g., For loop, While loop, File I/O, and so forth.
Signal processing	Collection of functions used for Signal Processing. It includes wfm generation, wfm conditioning, wfm measurement, filters, transforms, windowing, and so forth.
User Libraries	By default this palette is empty. The user can create libraries and add them under this subpalette.

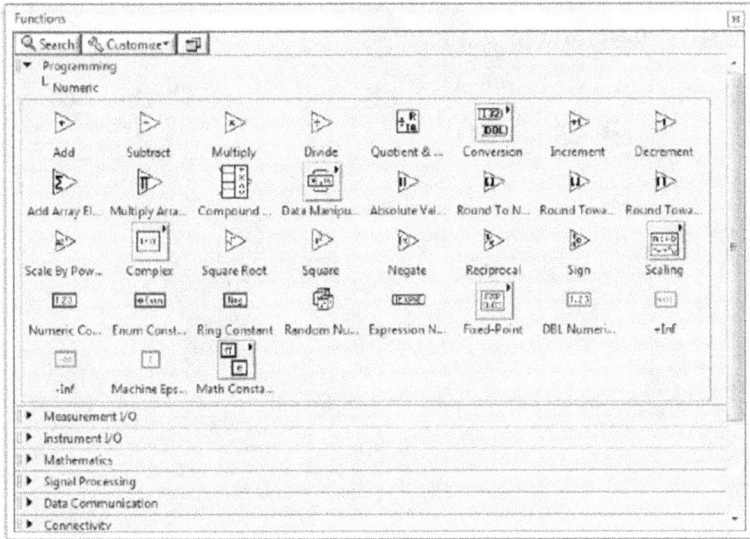

Figure 2.6. Functions palette.

environment, G, Functions Palette, and Controls Palette is helpful to build a sophisticated application.

However, if the user wants to skip the basics and cannot wait to get hands dirty with the nuts and bolts of LabVIEW, in this section we will build a simple calculator VI to get the feel of LabVIEW.

2.6.1 SIMPLE CALCULATOR

Steps to get a simple calculator up and running as follows:

1. Start LabVIEW.
2. Create project and save it as Calculator.vi.
3. Go to Block Diagram window. Right click on the workspace to open the Controls palette.
4. Under Modern > Numeric subpalette, find Numeric Control and Numeric Indicator. Drag two instances of Numeric Controls and four instances of Numeric Indicators. Arrange all the controls as shown in Figure 2.7.
5. Double click on the labels to edit the name of the control. Name one of the Numeric Indicators Num1 and the other Num2. Name the Numeric Indicators as "Num1 + Num2,""Num1 − Num2,""Num1 * Num2," and "Num1 / Num2," as shown in Figure 2.7.

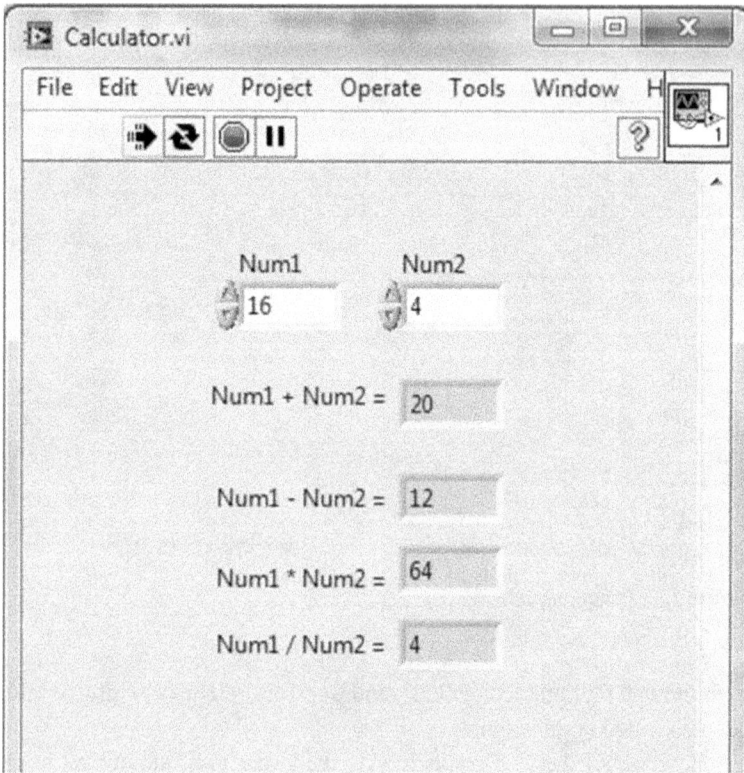

Figure 2.7. Front Panel of simple calculator.

6. Now go to the Block Diagram window.
7. Right click anywhere on the Block Diagram window, to bring up the Functions Palette.
8. Under Programming > Numeric category, find the Add, Subtract, Multiply, and Divide functions. Drag each of them onto the Block Diagram.
9. If the Automatic Tool Selection mode is not on, please turn it on as explained in the Tools Palette section. Once the "Automatic Tool Selection" mode is turned on, bring the mouse cursor closer to the terminals and watch the cursor change to Wiring Tool, automatically.
10. In the Wiring Tool mode, simply single click to start the wiring. Move the cursor all along toward the terminal we want to connect to. The wire will autoshape itself around the blocks. When the cursor is on the other terminal, single click again to end the wiring. Now continue with the wiring and complete the connections as shown in Figure 2.8.

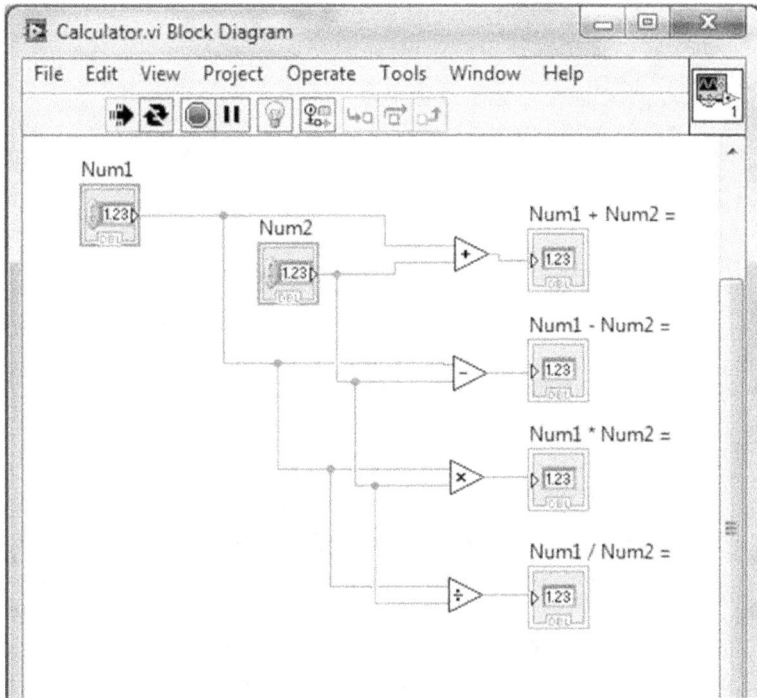

Figure 2.8. Block Diagram of simple calculator.

Table 2.5. Divide by zero behavior of LabVIEW

Num1	Num2	Num1/Num2	Num1 * Num2
0	0	NaN	0
1	0	Inf	0
−1	0	-Inf	-0

11. Once all the connections are in place, all you need to do is click on the "Run Continuously" button. You can now play around by inputting the values in the Numeric Controls. Please note that LabVIEW is smart enough to recognize the "Divide By Zero" case as shown in Table 2.5.

2.7   PROGRAMMING STRUCTURES

The available programming structures are shown in Figure 2.9. We will touch base with only a few of those structures.

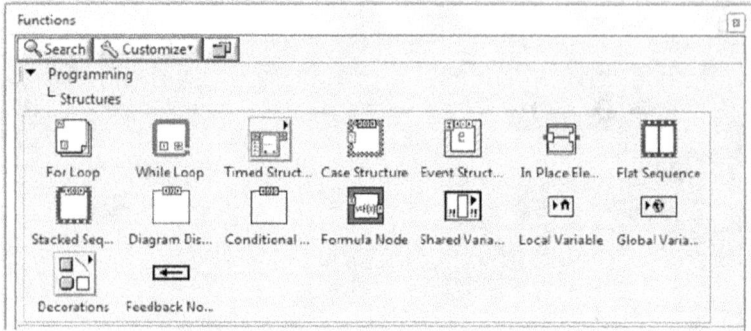

Figure 2.9. Programming structures in LabVIEW.

2.7.1 FOR LOOP

For loop executes a subdiagram $\boxed{N}$ number of times. $\boxed{N}$ is an input to the For Loop. $\boxed{i}$ is an output terminal of the For Loop. It indicates the completed iteration count. Let us design a simple Summation.vi to explain the "For Loop."

The Front Panel of Summation. vi is very simple with one Numeric Control and two Numeric Indicators as shown in Figure 2.10.

We have calculated the sum in two ways:

1. Using the For Loop
2. Using the formula $\sum n = \dfrac{n(n+1)}{2}$

As can be seen from the Block Diagram of the Summation.vi in Figure 2.11, the top section calculates the sum using For Loop and the bottom section calculates the sum using formula.

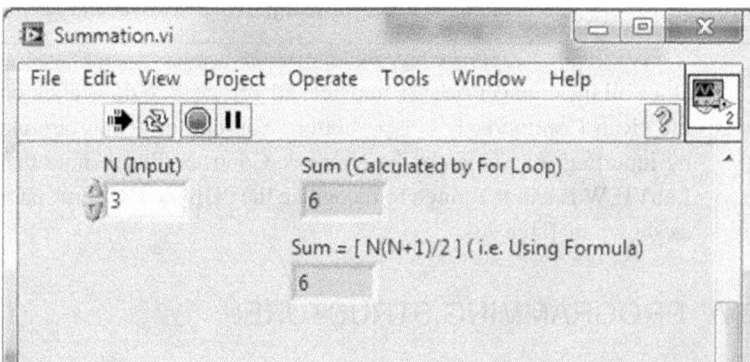

Figure 2.10. Example run of VI to demonstrate "For Loop".

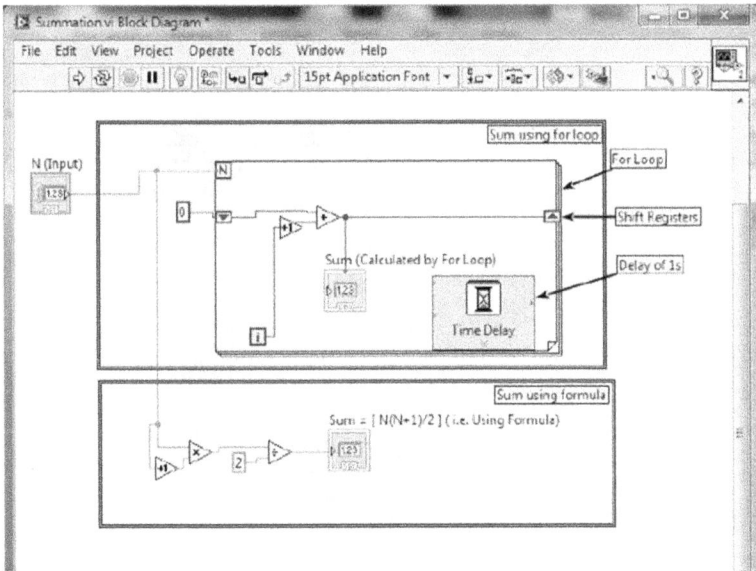

Figure 2.11. Block Diagram of VI to demonstrate the "For Loop".

The time delay of 1 s allows the user to see the Numeric Indicator named "Sum (Calculated by For Loop)" and update the sum on every iteration. It helps the user appreciate the fact that the sum is calculated by going over the Add operation N number of times.

Shift Register is a new control in this VI. It is used for passing the values between iterations of the loop. We pass the sum calculated in the current iteration to the next iteration through the Shift Register. The Shift Register can be introduced to a loop structure by right-clicking the left or right border of a loop and selecting the Add Shift Register from the short-cut menu. We need to initialize the shift registers otherwise; it uses the old value from the last run as its default value. In our case, we initialized the shift register to zero by wiring a constant 0 to the input terminal of the shift register ⓪—▾.

2.7.2 SEQUENCE

In some cases the order of execution is of importance. We want to maintain a dependency of events. For example, we want to save the readings of an experiment to a file only after all the readings are available. Once the file is saved to a disk, only then do we want to upload the file to say an FTP server. This is where the sequence structure comes in handy.

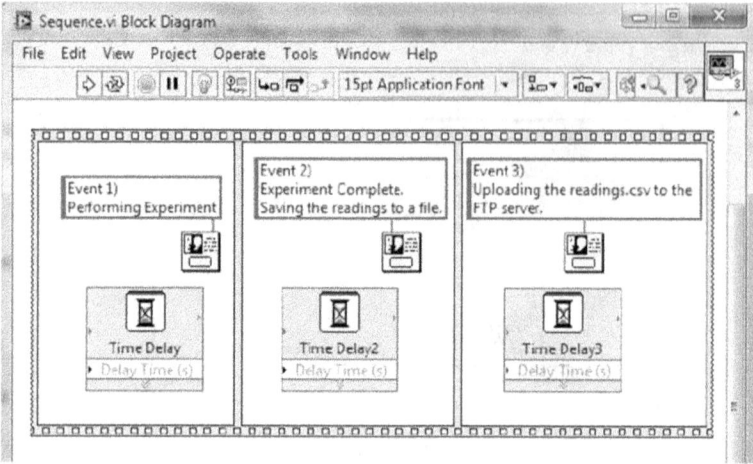

Figure 2.12. Block Diagram of a VI to demonstrate "sequence".

Figure 2.13. Example run of a VI.

Sequence structure contains a set of subdiagrams or frames that execute in sequence. There are two types of Sequence structures, namely, flat sequence and stacked sequence. The only difference is in their view. A stacked sequence shows only one frame at a time as it has all the frames in a stack. It saves space on the Block Diagram. A flat sequence is shown in the Block Diagram in Figure 2.12.

As shown in Figure 2.12, the flat sequence shows three frames. The sequence starts from left to right. The message boxes appear in the sequential order, as shown in Figure 2.13.

2.7.3 CASE

Case structure executes only one case out of the available cases based on the selector value arrived at the selector terminal. The VI of the V-Lab experiment makes use of the case structure to choose the subdiagram based on the mode of operation, namely, co-current or concurrent.

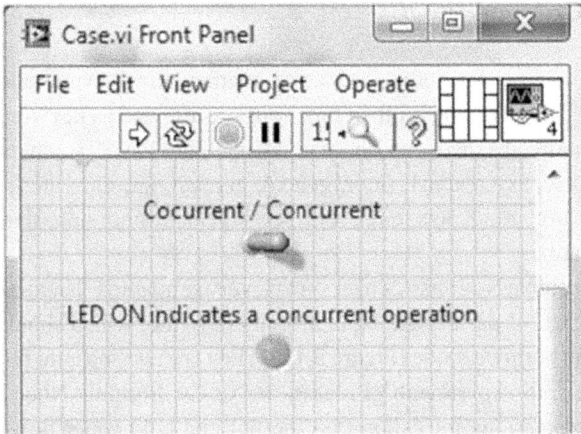

Figure 2.14. Front Panel of a VI to demonstrate "case structure".

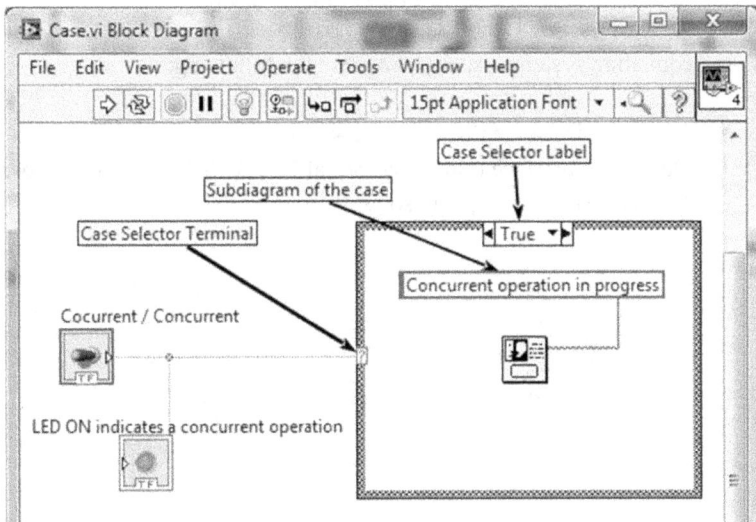

Figure 2.15. Block Diagram of a VI to demonstrate "Case Structure".

As can be seen, the toggle button on the Front Panel in Figure 2.14 is wired to the selector input terminal of the case structure in Figure 2.15. In our example the selector value is of type Boolean. Similarly we can tie String, Integer, enumerated type, and error cluster to the Selector terminal. The Selector Label shows the subdiagram for that particular case selector value.

2.7.4 FORMULA NODE

Formula Node is used for evaluating mathematical expressions (Figure 2.16). The following built-in functions are allowed as keywords in the Formula node: abs, acos, acosh, asin, asinh, atan, atan2, atanh, ceil, cos, cosh, cot, csc, exp, expm1, floor, getexp, getman, int, intrz, ln, lnp1, log, log2, max, min, mod, pow, rand, rem, sec, sign, sin, sinc, sinh, sizeOfDim, sqrt, tan, tanh.

The Formula Node has its own syntax, very similar to C language. There are mathematical parsers available in LabVIEW too, which should not be confused as Formula Node (Figure 2.17). They are two separate functions.

The V-Lab experiment VI makes use of the Formula Node to evaluate the expressions for calibrating the hot and cold water flows during runtime.

2.8 DATA ACQUISITION WITH LABVIEW

Data acquisition systems usually comprise the following:

• Device or instrument under observation
• Hardware interface (Universal Serial Bus [USB], General Purpose Interface Bus [GPIB], and so on.)

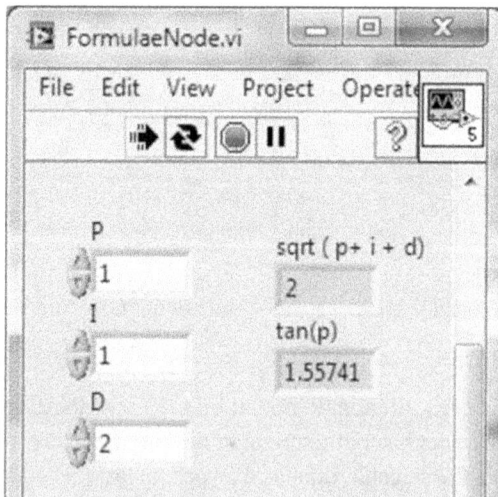

Figure 2.16. Front Panel of a VI to demonstrate "Formula Node".

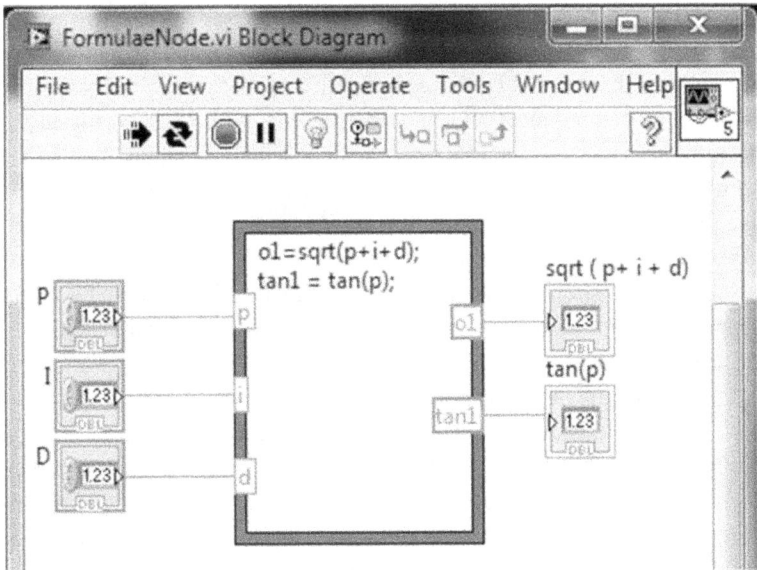

Figure 2.17. Block Diagram of a VI to demonstrate "Functions Node".

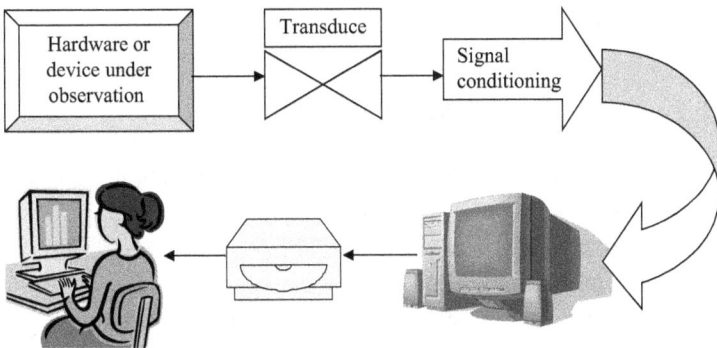

Figure 2.18. Data acquisition system components.

- Computer
- Driver software
- Application software (Figure 2.18)

2.8.1 VIRTUAL INSTRUMENTATION

An instrument is a device that measures a physical quantity such as temperature, flow, pressure, and so forth, and the term "instrumentation" simply means putting together such instruments in order to measure and control the process variables. And virtual instrumentation comprises instruments that are not hardware but software that mimic the hardware. A physical device called analog to digital converter (ADC) is used for converting the analog signals to digital signals before feeding them to the computers.

A synthetic instrument is a virtual instrument that is used for specific synthesis, analysis, or measurement purposes. The term "synthetic" in "synthetic instrument" is a kind of misnomer in the sense that it might seem to imply that the synthetic instrument is particularly a synthesizer. Instead, because the instrument itself is synthesized by software, it is called so. The instrument can be a pure analyzer or even a synthesizer.

A synthetic instrument working group (SIWG) has been created by DoD (Dept. of Defense, USA) for the sole purpose of defining standards for interoperability of synthetic instrument systems. SIWG defines synthetic instrument (SI) as "A reconfigurable system that links a series of elemental hardware and software components with standardized interfaces to generate signals or make measurements using numeric processing techniques."

Virtualization drastically reduces the cost of and boosts the speed of instrumentation.

2.8.2 COMMUNICATION WITH DAQ DEVICES

The data acquisition (DAQ) system usually consists of a transducer or a sensor, ADC, driver software, and application software.

A transducer is a sensor that is used to sense the physical quantity and transform it into an electrical signal. The electrical signal is either a voltage signal or a current signal. The signal obtained from a sensor is usually of a very low magnitude. Hence, an amplifier is used to increase the magnitude of the signal in a proportionate manner without losing the information in the signal. This amplified signal is then passed through the ADC circuit for digitizing the signal.

The ADC circuit consists of components such as Sample and Hold Circuit, Quantizer, and Encoder (Figure 2.19). As the name suggests, Sample and Hold circuit simply samples a signal and holds it for a minimum period of time. It eliminates the variations in the input signal that can corrupt the quantization process. The quantization process maps the continuously varying input signal to a finite set of values, that is, rounding the value to given precision. In the process, the quantizer is bound to introduce a round-off-error due to truncation. This error is called as

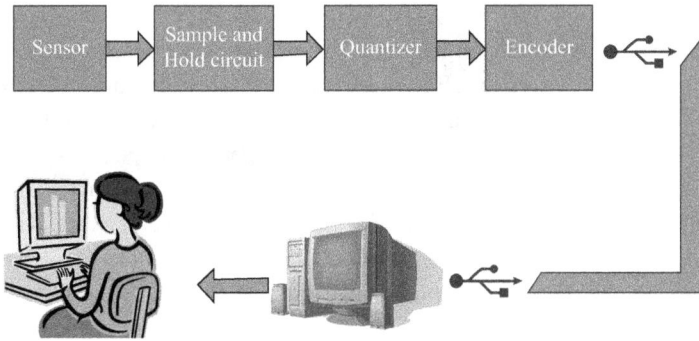

Figure 2.19. Dataflow in DAQ system.

quantization error. A quantized signal is then passed to the encoder. The encoder encodes the signal in order to correct and reduce the transmission errors. The encoded signal is then passed over to the computer over the USB or GPIB interface.

The driver software is specific to its hardware. It exposes a set of function calls that facilitate communication with the hardware. The application software builds on top of the driver. It calls the functions and passes the parameters through these function calls to the device. For example:

a. **function** ReadAnalog (**var** channel: integer; **var** value: integer);
b. **procedure** WriteAnalogs (**var** AO1: integer; **var** AO2: integer); **stdcall;**
c. **procedure** ReadDigital (**var** channel: integer; **var** value: integer); **stdcall;**
d. **procedure** WriteDigitals (**var** DO1: integer; **var** DO2: integer; **var** DO3: integer; **var** DO4: integer; **var** DO5: integer; **var** DO6: integer; **var** DO7: integer; **var** DO8: integer); **stdcall;**
e. **procedure** SetMode (**var** Mode: integer); **stdcall;**

2.8.3   GENERAL PURPOSE INTERFACE BUS

Originally it was developed by HP in the 1960s as an easier way to interface the instruments and controllers for automated test equipment. It was called Hewlett-Packard Interface Bus (HP-IB). Other companies began using it and christened it the GPIB (general purpose interface bus). GPIB was standardized in 1975 by the IEEE and again in 1978 and 1987. IEEE 448.1 became the standard for the connector, and IEEE 488.2 the control command standard. GPIB is a well-adopted standard for connecting fast as well as slow devices in the same system. Up to 15 connectors can be stacked together. In LabVIEW, there are tools called Instrument Drivers, which are collections of functions and VIs that implement the commands necessary to perform the GPIB instrument's operations.

The flip side of GPIB is that the connector is large and bulky. USB is a cheaper alternative to GPIB. Besides, GPIB can deliver the data at maximum transfer rate of 8 Mbits/s whereas USB 3.0 can transfer at 5 Gbits/s. Every general purpose computer comes with USB ports. It is not the case with GPIB.

2.8.4 UNIVERSAL SERIAL BUS

Universal serial bus was originally designed in the mid-1990s to standardize the connection of computer peripherals such as keyboard, mouse, printer, and so forth. USB has almost kicked out serial and parallel ports from the general computer world. Nowadays, many portable devices

Type	Port image	Connector image
Type A	4.5 mm × 12.0 mm 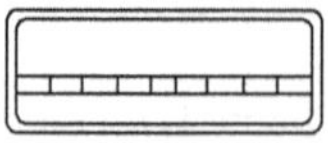	
Type B	7.3 mm × 8.5 mm 	
Mini-A	3.0 mm × 6.8 mm 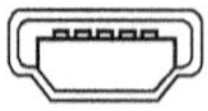	
Mini-B	3.0 mm × 6.8 mm 	

Figure 2.20. USB ports.

come with USB power chargers. Also, the instrumentation world banks upon this cheaper and faster interface standard. USB defines four types of ports as shown in Figure 2.20.

The Armfield's heat exchanger equipment is fitted with USB interface with Type-B port. A USB cable with Type-A and Type-B port is also provided with the setup. Hence, it is very easy to connect the device to any general purpose computer that has a USB port.

2.9 QUESTIONS

1. **What is G language? What are the advantages of G language over textual languages?**
   The answer to this question can be found in Section 2.2 G-language.
2. **How to get an evaluation copy of LabVIEW?**
   https://lumen.ni.com/nicif/us/cdlveval/content.xhtml
3. **Can the evaluation period be extended?**
   Yes, it can be extended by 45 days. Extension is allowed only once.
   https://delta.ni.com/extendedevaluation/index.xhtml?p_lang=US

CHAPTER 3

HARDWARE:
ARMFIELD HEAT EXCHANGER
AND HT30XC SERVICE UNIT

This chapter describes the hardware we used for the Heat Exchanger Virtual Laboratory. The hardware used was from the Armfield company. We will discuss the details of how to interface a computer with this hardware. Following details are provided in this chapter:

• Details about channel allocations,
• Significance of each channel and its signal,
• Universal serial bus interface function calls.

3.1 OPERATING HT30XC USING CUSTOMER-GENERATED SOFTWARE

We have used HT30XC for this project (Figure 3.1).

3.1.1 CHANNEL ALLOCATIONS

The interface between the Armfield heat exchanger bench and the computer is a universal serial bus (USB) interface, meeting the standard Microsoft protocols. Armfield is registered with Microsoft as an authorized supplier of USB interfacing equipment.

The interface is capable of passing data on 26 channels, as described in the following text:

1. Analog Inputs: 8 differential channels or 16 single-ended channels, each with –5V to 5V signals digitized into a 12-bit number. The interface will pass a value between –2047 and 2047 to the computer.

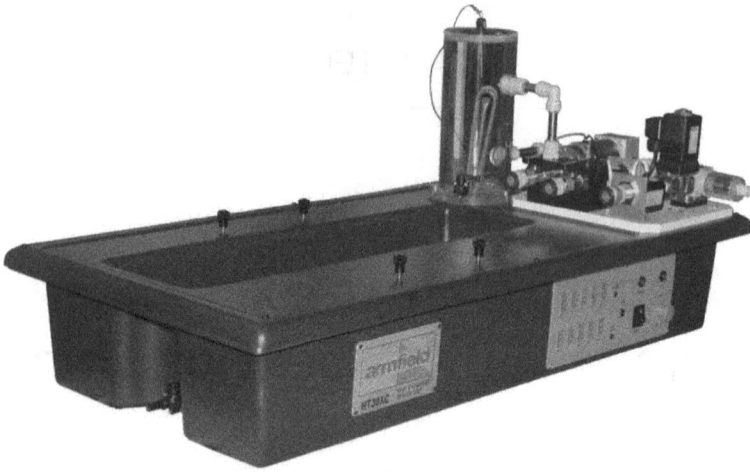

Figure 3.1. Armfield Heat Exchanger service unit. (*Source*: HT30XC Instruction Manual.)

2. Analog Outputs: Two channels, each with –5V to 5V signals, taken from a 12-bit number. Computer must pass a value between –2047 and 2047 to the unit.
3. Digital Inputs: Eight channels each receiving a 0 or 1.
4. Digital Outputs: Eight channels each passing a 0 or 1.

The channel allocations for the HT30XC are tabulated in the following text.

Table 0.1. Analog Signals from Heat Exchanger to Computer (*source*: HT30XC User Guide)

Channel	Code	Use	Scaling
0	T1	Hot water temperature	–5V = 0°C, +5V = 133°C
1	T2	Hot water temperature	–5V = 0°C, +5V = 133°C
2	T3	Hot water temperature	–5V = 0°C, +5V = 133°C
3	T4	Hot water temperature	–5V = 0°C, +5V = 133°C
4	T5	Hot water temperature	–5V = 0°C, +5V = 133°C
5	T6	Cold water temperature	–5V = 0°C, +5V = 133°C
6	T7	Cold water temperature	–5V = 0°C, +5V = 133°C
7	T8	Cold water temperature	–5V = 0°C, +5V = 133°C
8	T9	Cold water temperature	–5V = 0°C, +5V = 133°C

9	T10	Cold water temperature	$-5V = 0°C, +5V = 133°C$
10	F1	Hot water flow	$0V = 0$ L/min, $5V = 25$ L/min
11	F2	Cold water flow	$0V = 0$ L/min, $5V = 5$ L/min
12	Not used		
13	Not used		
14	Not used		
15	Not used		

Table 0.2. Analog Signals from Computer to Process (*source*: HT30XC User Guide)

Channel	Code	Description	Scaling
0	P1	Hot water pump speed	$0V$ = stopped, $5V$ = full speed
1	V1	Cold water valve setting	$0V$ = Closed, $1.5V$ = Just Opening, $3.5V$ = Fully open

Table 0.3. Digital Signals from Process to Computer (*source*: HT30XC User Guide)

Channel	Title	Description	Scaling
0		Not used	
1		Not used	
2	Low level	Monitors the water level in the hot water vessel	0 = Low Level; 1 = OK
3		Not used	
4	Thermostat/ level monitor	Monitors the output of the over-temperature thermostat AND the water level in the hot water vessel	0 = Over Temp; 1 = OK
5		Not used	
6		Not used	
7		Not used	

Table 0.4. Digital Signals from computer to process (*source*: HT30XC User Guide)

Channel	Title	Description	Scaling
0	Power on request	Allows the power on relay to be energized, subject to the presence of an appropriate watchdog pulse	0 = Power Off; 1 = Power On
1	Watchdog pulse	Pulsed signal to keep the watchdog circuit energized, enabling the heat exchanger bench power to be turned on	Pulsed signal, min rate 1 pulse every 5 seconds
2	SSR drive	Time-modulated signal controlling the hot water heater solid state relay (SSR)	0 = heater off; 1 = heater on
3	Pump direction	Controls the change over relay that reverses the hot water pump direction	0 = countercurrent; 1 = co-current
4	Stirrer on	Only used on HT34	0 = off; 1 = on
5	Aux.heater control		0 = off; 1 = on
6		Not used	
7		Not used	

3.2 USB INTERFACE DRIVER FUNCTION CALLS

Armfield Heat Exchanger comes with a driver file ARMIFD.DLL. This interface driver DLL file exposes functions for four types of data I/O as explained in the following text. The integer is a 32-bit integer type.

3.2.1 READ ANALOG

This function takes the channel number as its argument and returns the current value of physical quantity at that channel.

function Read Analog (**var** channel: integer; **var** value: integer); **stdcall;**

Table 3.1. Channels and their significance

Channels	Description
0–7	Differential channels
0–15	Single-ended channels
16–31	Multiplexed channels

Table 3.2. Significance of values returned by ReadAnalog function

Returned value	Description
±2047	±5V
9999	Error

The std call directive indicates that the call is handled in a way that is recognizable by most programming languages, including LabVIEW. Table 0-4 shows how the channel number should be selected:

Table 3.2 shows the meaning of the values returned.

3.2.2 WRITE ANALOG

procedure WriteAnalogs (**var** AO1: integer; **var** AO2: integer); **stdcall;**

There are only two analog channels as explained in Table 0.2. This call sends values to both the analog output channels. The values sent should be between ±2047 corresponding to ±5V.

3.2.3 READ DIGITAL

procedure ReadDigital (**var** channel: integer; **var** value: integer); **stdcall;**

This call returns the values from one of the eight digital channels mentioned in Table 0.3. The meaning of the returned values is described in Table 3.3.

3.2.4 WRITE DIGITAL

procedure WriteDigitals (**var**DO1: integer; **var**DO2: integer; **var**DO3 : integer; **var**DO4 : integer; **var**DO5 : integer; **var**DO6 : integer; **var**DO7 : integer; **var**DO8 : integer); **stdcall;**

Table 3.3. Significance of values returned by ReadDigital function

Returned value	Description
0	Channel is off
1	Channel is on
−1	Error

Table 3.4. Significance of "mode" argument in SetMode function

Mode value	Description
0	8 differential channels
1	16 open-ended channels

This call writes values from the eight digital output channels. Please note that all the arguments to this call have to be either 0 or 1.

3.2.5 SET MODE

    procedure SetMode (**var**Mode : integer); **stdcall**;

One needs to set the device to single-ended or differential mode. This function allows us to do so. Table 3.4 explains the meaning of the argument mode.

3.3 LABVIEW DATA LOGGER

3.3.1 READANALOG INPUT CHANNELS

1. Go to the Block Diagram window and right click to open the Functions palette.
2. From the functions palette, pick "Call Library Function" and place it on the block diagram. Search for Call Library Function or directly find it under Connectivity > Libraries & Executables subpalette as shown in the Figure 3.2.
3. Double click the "Call Library Function" icon to open the settings UI.

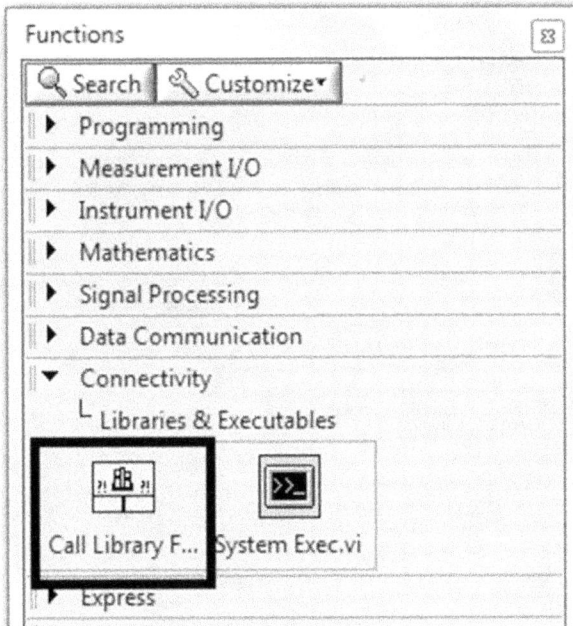

Figure 3.2. Location of "Call Library Function" VI.

4. As shown in Figure 3.3, go to the "Function" tab. Type the path of the DLL in "Library name or path" textbox or use the browser to find the path and select it.
5. Type function name as ReadAnalog. Alternatively, if the ArmIFD.dll path is valid, then choose the function from the dropdown.
6. Choose "Calling convention" as "stdcall (WINAPI)."
7. Choose "Thread" as "Run in UI thread."
8. Go to the "Parameters" tab and add two parameters of the type Signed 32-bit Integer, namely, channel and value. We need to "Pass" the parameters as "Pointer to Value." Also set the data type of "return type" parameter as "void" as shown in Figure 3.4.

9. Notice that the icon has changed to .
10. Note that there are two input and two output terminals to the icon.
11. Wire a Numeric Control to the input named channel. And wire a Numeric Indicator to output named value.
12. Run the VI. See the value that is being displayed in the Numeric Indicator named Value.

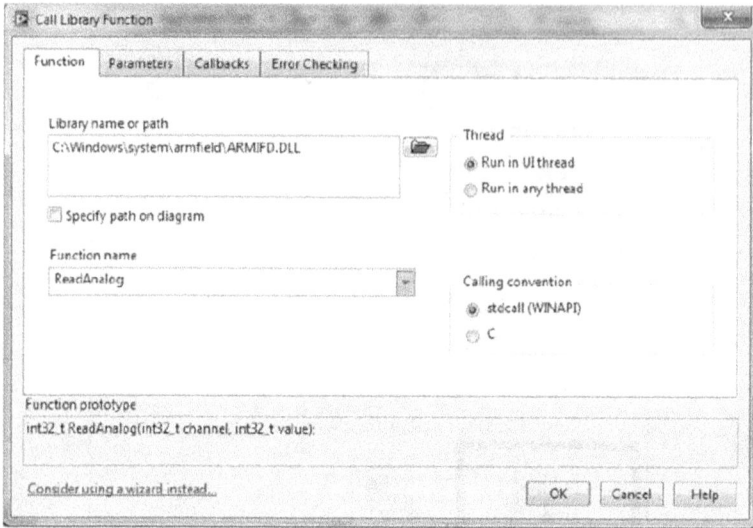

Figure 3.3. Function tab setup of "ReadAnalog" function via "Call Library Function" VI

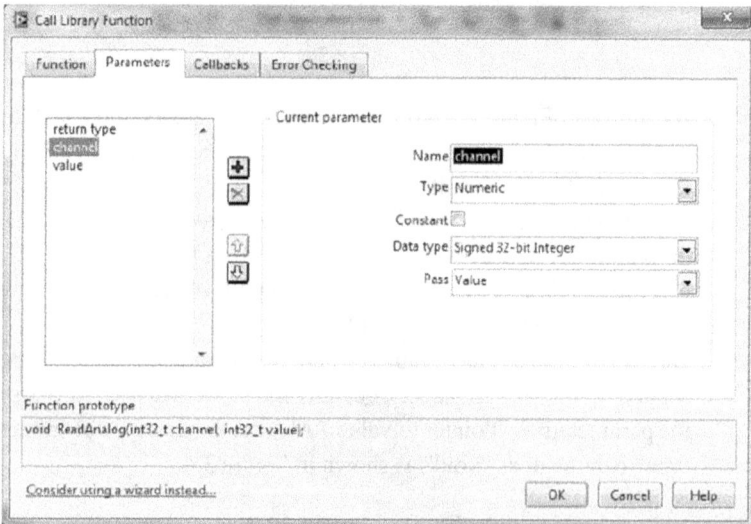

Figure 3.4. Parameters tab setup of "ReadAnalog" function via "Call Library Function" VI.

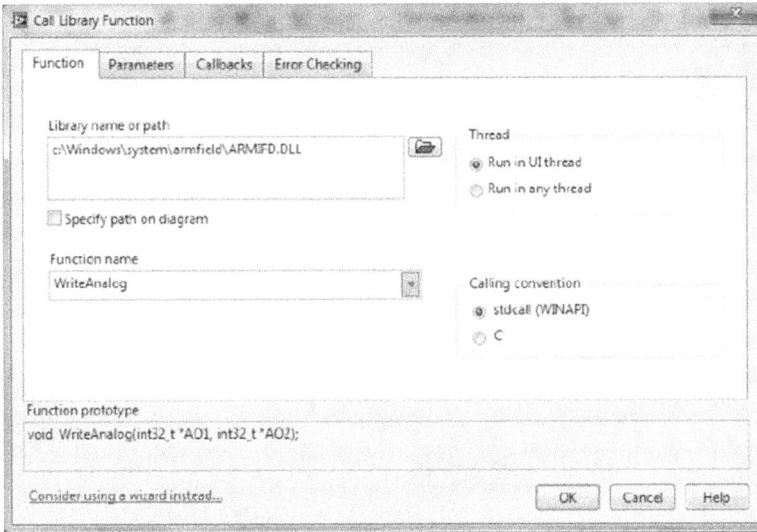

Figure 3.5. Function tab setup of "WriteAnalog" function via "Call Library Function" VI

3.3.2 READ THE DIGITAL INPUT CHANNELS

Follow the same steps as explained in Section 3.1.3.1 except that the function name to be specified is ReadDigital.

3.3.3 WRITING THE ANALOG OUTPUTS

Figure 3.5 shows Call Library Function setup for writing to analog channels. Please make sure that the controls tied to the two input terminals are clamped at -2047 and +2047. This can be done through the Data Entry tab of the Properties window for the Numeric Control. Set Response to value outside limit to Coerce.

Also, set the data representation to *Word* (I16) as shown in Figure 3.6 and Figure 3.7.

3.3.4 WRITING THE DIGITAL OUTPUTS

Follow the same steps as explained in Section 3.1.3.1 except that the function name to be specified is WriteDigital and add eight parameters, named DO1 to DO8 to this function.

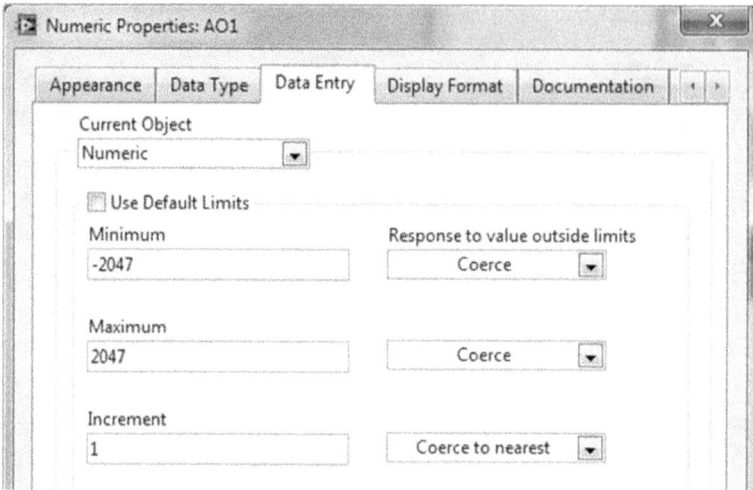

Figure 3.6. Data representation for the arguments of WriteAnalog function

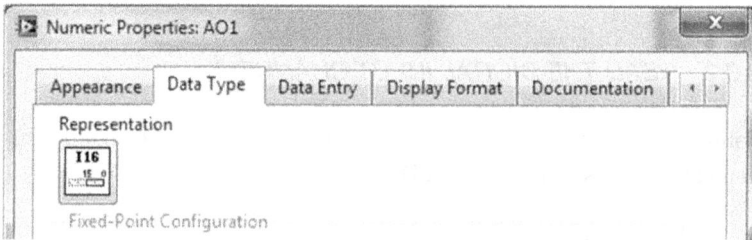

Figure 3.7. Selection of data representation as I16

CHAPTER 4

DESIGN OF LABVIEW VI PROGRAM

This chapter details all the aspects of the LabVIEW VI program, used for the Heat Exchanger experiment. Every control used and its purpose in the virtual instrument (VI) is explained. It gives a fair idea about how the VI was built. Hence, it gives the reader enough material to start building one of his or her own for future work.

Once the knowledge of how to use the LabVIEW controls and indicators is obtained, as mentioned earlier, LabVIEW programming is as easy as dragging and dropping the required controls and indicators—very much similar to putting together puzzle pieces! One gets to experience real rapid prototype development in the world of instrumentation, by virtue of LabVIEW. The ability of rapid prototyping in turn helps the designer incrementally better the design.

Let us begin with the algorithm of the VI.

4.1 SOFTWARE: ALGORITHM OF THE PROGRAM

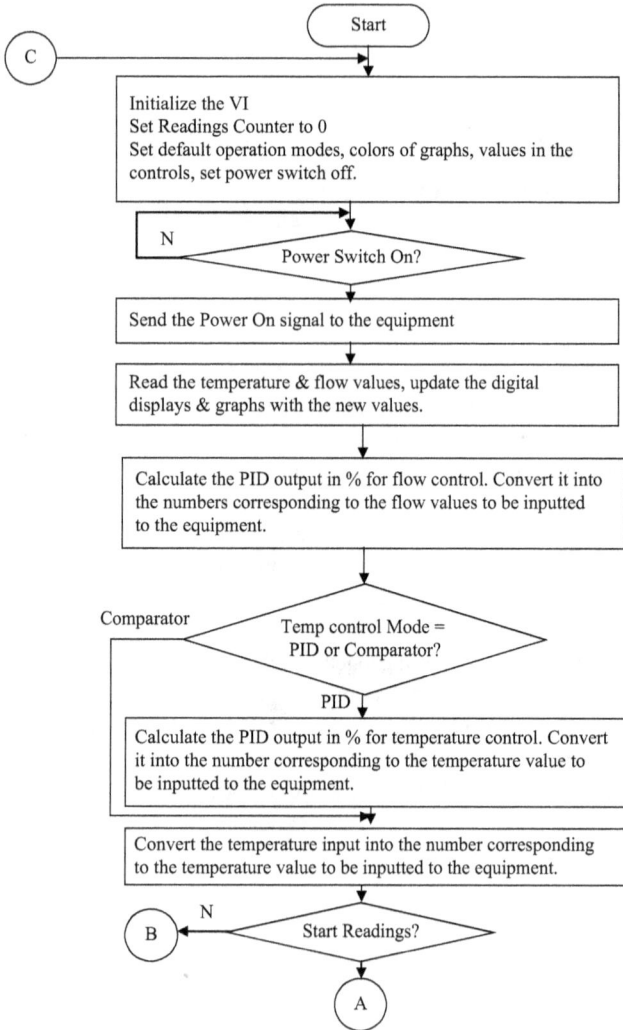

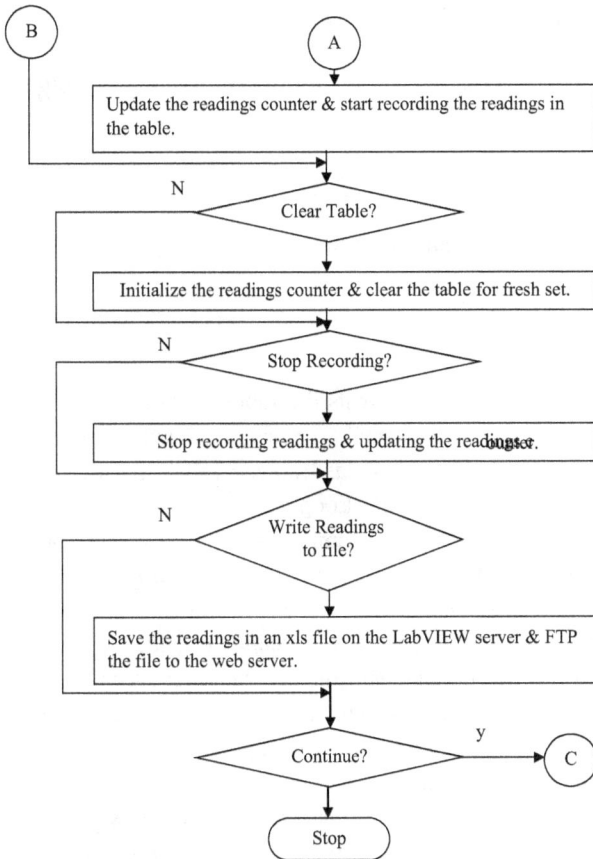

4.2 INTRODUCTION OF LABVIEW CONTROLS USED IN THE PROJECT

Though it is not a prerequisite, it is recommended to read Chapter 2 "Lab-VIEW Basics before starting with this section". It will help to get a background about LabVIEW.

The controls used in the project can be broadly classified into following categories:

- Execution controls: They are explained in brief in the following text.
- Other block diagram objects: They are explained as and when they are encountered in the front panel and block diagram design in the following text.

4.2.1 EXECUTION CONTROLS

There are three execution controls that were used extensively in the project, namely:

1. While loop
2. Case structure
3. The stacked sequence structure

4.2.1.1 While Loop

It repeats the subdiagram inside it until the conditional terminal, an input terminal, receives a particular Boolean value, that is, in our case the While Loop executes indefinitely until the VI is in the running mode. The While Loop is so configured that it executes at least once. The iteration [i] terminal provides the current loop iteration count, which is zero for the first iteration.

The hardware HT30XC has a watchdog circuit. This circuit acts as a safety valve. It expects a signal from the computer every 10 s. If a signal is not received, the circuit considers it as a computer crash or software failure. In such a case, it automatically switches off the heater and the pump thus keeping the hardware safe.

We have used the [i] terminal to generate the watchdog pulses. A minimum of 1 Watchdog pulse every 5 s is needed by Armfield's equipment in order to run the equipment. Our LabVIEW interface provides a watchdog pulse to the equipment every other second (1 pulse/2 s).

4.2.1.2 Case Structure

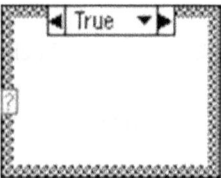

This structure has many cases, exactly one of which executes when the structure executes. The value wired to the selector terminal determines which case to execute. You can also label the case the way you want. The label can be a Boolean, string, integer, or enumerated. Right click the structure border to add or delete cases.

We have used this structure in five places in our block diagram:

1. To select the type of heater control
2. To execute the case depending upon whether the heater control is on or off

3. To execute the case when the user hits "Record Data"
4. To execute the case when the user hits "Clear Table"
5. To execute the case when the user hits "Write readings to file"

4.2.1.3 Stacked Sequence Structure

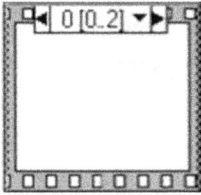

The Stacked Sequence structure, shown on the left, stacks each frame so we see only one frame at a time and executes frame 0, then frame 1, and so on until the last frame executes. The Stacked Sequence structure returns data only after the last frame executes. The Stacked Sequence structure helps to conserve space on the block diagram.

We have made use of this structure to implement a simple state machine with two states. In first state we store the recorded readings from the LabVIEW interface into an excel file onto the hard drive of the lab's computer. In the second state, we "FTP" it to the server. The server is used to store the latest READINGS.xls file, which then can be downloaded using a hyperlink provided on the web page of the experiment.

4.3 DESIGN OF FRONT PANEL

The front panel makes the web interface that the students use to perform the experiment online. The front panel consists of three tabs, namely, Pre-lab Instructions, Controls, and Readings. We will discuss each tab's design in detail in the following sections.

4.3.1 PRE-LAB INSTRUCTIONS TAB

Figure 4.1 shows the Pre-lab instructions. The same pre-lab instructions are shared as a part of experiment material. However, putting it up on the experiment UI, is a convenience for the student.

4.3.1.1 Architecture

This tab has nothing more than a "Text label control."

4.3.1.2 Intended Function

This tab contains all the instructions that an experimenter is expected to read and understand. They will facilitate the experimenter to use the interface.

4.3.2 CONTROLS TAB

The Controls Tab is the heart of this VI. All the controls and indicators are on this tab. The user/student uses this tab to perform the experiment. This tab can be vertically divided into two halves. The left half contains the controls whereas the right half consists of indicators and graphs. In the left half the relevant controls are separated into five logical groups. The right half contains the three graph plotters for hot water flow, cold water flow, and temperatures. Each of these graphs and the controlgroups are explained in brief in Table 4.1.

4.3.2.1 Architecture

The "Controls tab" (Figure 4.2) contains the following:

4.3.2.2 Intended Function

This is the main tab, which is used for performing the experiment. It contains all the controls and indicators needed. It can be used to input various parameters and observe the real time graphs of hot flow rate (L/min), cold flow rate (L/min), and temperature (°C).

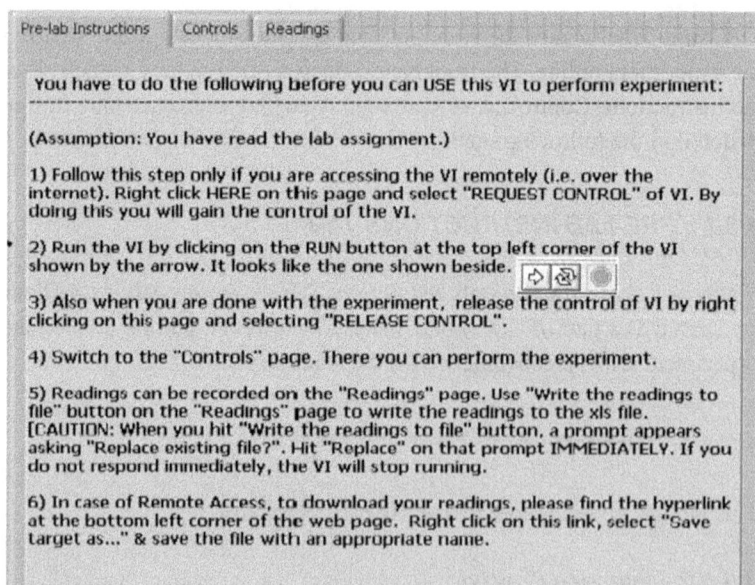

| Pre-lab Instructions | Controls | Readings |

You have to do the following before you can USE this VI to perform experiment:

(Assumption: You have read the lab assignment.)

1) Follow this step only if you are accessing the VI remotely (i.e. over the internet). Right click HERE on this page and select "REQUEST CONTROL" of VI. By doing this you will gain the control of the VI.

2) Run the VI by clicking on the RUN button at the top left corner of the VI shown by the arrow. It looks like the one shown beside.

3) Also when you are done with the experiment, release the control of VI by right clicking on this page and selecting "RELEASE CONTROL".

4) Switch to the "Controls" page. There you can perform the experiment.

5) Readings can be recorded on the "Readings" page. Use "Write the readings to file" button on the "Readings" page to write the readings to the xls file.
[CAUTION: When you hit "Write the readings to file" button, a prompt appears asking "Replace existing file?". Hit "Replace" on that prompt IMMEDIATELY. If you do not respond immediately, the VI will stop running.

6) In case of Remote Access, to download your readings, please find the hyperlink at the bottom left corner of the web page. Right click on this link, select "Save target as..." & save the file with an appropriate name.

Figure 4.1. Pre-lab instructions tab.

Table 4.1. Architectural components of "Controls Tab"

Snapshot of the Controls from "Controls tab"	Description
Indicator for Low Water Level in the Hot water tank.	When the water level in the hot water tank goes below the sensor's level, this indicator turns RED, and the equipment stops running.
Mode of Operation — Cocurrent / Countercurrent	*Co-current Mode:* Hot and cold water flow in the same direction through the shell and tube heat exchanger. *Countercurrent Mode:* Hot water flows in the opposite direction of cold water through the shell and tube heat exchanger.
Power On Switch	This is a power switch, when turned on, an LED glows on the equipment that brings the equipment to life.
Hot water flow rate 3 lit/min / Cold water flow rate 1 lit/min / Temperature Setpoint 32 Degree Celcius	Using these controls the flow rates and temperature can be set. Hot water flow rate range: 0 to 5 L/min Cold water flow rate range: 0 to 5 L/min Temperature set point range: 0 to 100°C

This switch is used to switch between two control modes, namely, proportional-integral-derivative Control and Comparator Control.

This cluster of text control is used to set the P, I, and D parameter values.

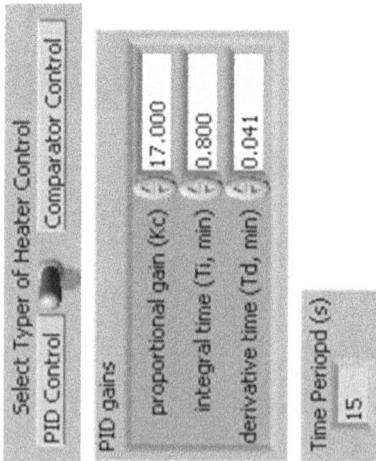

This is an indicator that shows the current time period selected for the pulse width modulation (PWM).

This is a simple picture control that is used to describe the schematic of the shell and heat exchanger. It also shows the thermostat locations with four dynamically selectable colors.

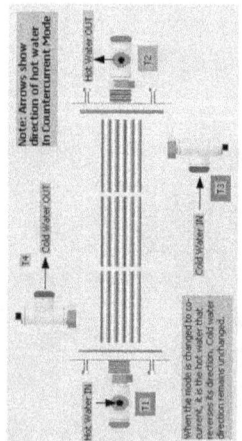

This is a graph control that plots the four temperature graphs. Each graph bears the corresponding color. The graph color can be changed dynamically. Since there are four graphs displayed at a time, this dynamic color selection feature comes handy in correlating each graph to its thermostat location on schematic and the digital reading beside.

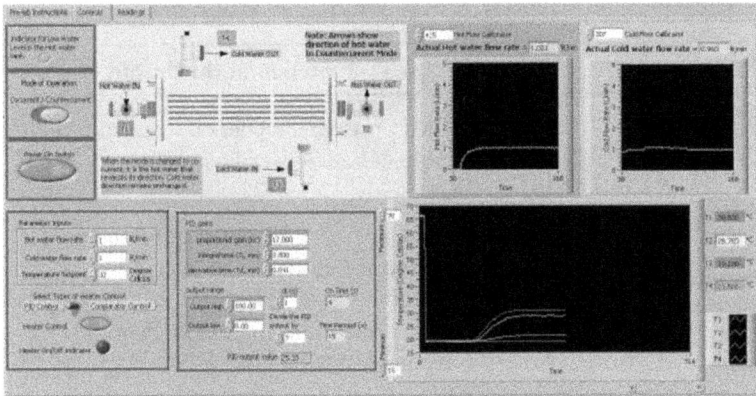

Figure 4.2. Controls tab.

4.3.3 READINGS TAB

Figure 4.3. Readings tab.

4.3.3.1 Architecture

This tab contains three push buttons and a table (Table 4.2).

4.3.3.2 Intended Function

This tab is where the experimenter can record the readings while the VI is running. The readings can be written or exported to an excel file. If needed, the table can be cleared to start a fresh set of readings.

Table 4.2. Functional description of the push buttons on "Readings Tab"

Push button	Function of the button
Record data	When pressed, it records the readings in the table. It keeps recording until depressed explicitly.
Clear table	In order to clear the table, one first needs to stop recording, if the "Record Data" button was pressed earlier. And then press the "Clear Table" button. It clears all the readings in the table.
Write the readings to file	It writes the readings from the table to an excel file named readings. xls and saves it on the C:\ of the server. When this push button is pressed, it pops up a message asking "Replace an existing file?" One needs to hit the "Replace" button on that message immediately. Otherwise, the VI stops running until the "Replace" button is hit.

4.4 DESIGN OF BLOCK DIAGRAM

The Block Diagram contains the entire LabVIEW program. Figure 4.4 shows an analogy, where the Block Diagram is the "back office operation center or a factory" and the Front Panel is the "Front Desk." The way the "Front Desk staff" liaises between customers and the "Back Office," Lab-VIEW "Front Panel" accepts the inputs from the user and communicates the outputs to the user in a presentable manner, while the actual processing of the inputs is taken care of by the Block Diagram.

Figure 4.5 shows the block diagram used by us to implement the virtual Heat Exchanger lab. The figure at the first glance appears to be very complex. However, LabVIEW's strengths abstracted the complexity away from the prototyping process and it appeared to be easier while designing this complex looking VI.

Explaining every control on the VI is out of scope. Instead, it is better that the VI be downloaded from the publisher's website and self-studied.

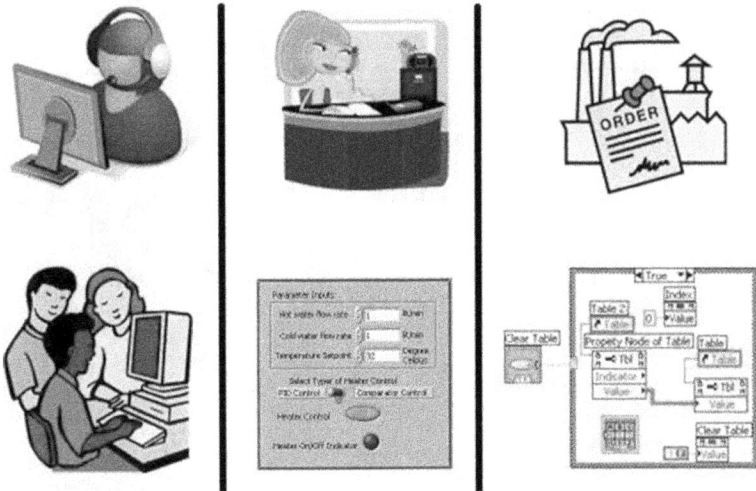

Figure 4.4. Analogy of Block Diagram and Front Panel.

4.5 HOW WERE THE PID PARAMETERS' VALUES DERIVED FOR TEMPERATURE CONTROL?

There are two approaches:

1. Analytical (mathematical derivation): The analytical approach is followed when the process is not available and the PID parameters need to be found before the design. In this one needs to have as accurate a mathematical model of the system as possible. This method is explained in Section 4.5.1.
2. Experimental (the ultimate cycle method): However, in our case, we had the process available to experiment with. In such a case, where the system is available to experiment, there are two methods to find out the optimum PID parameters, namely:
   1. The ultimate cycle method
   2. The process reaction method
   We used the ultimate cycle method that required us to get the two measurements from the process and use those measurements to get the PID parameter values using the formulae given by the method.

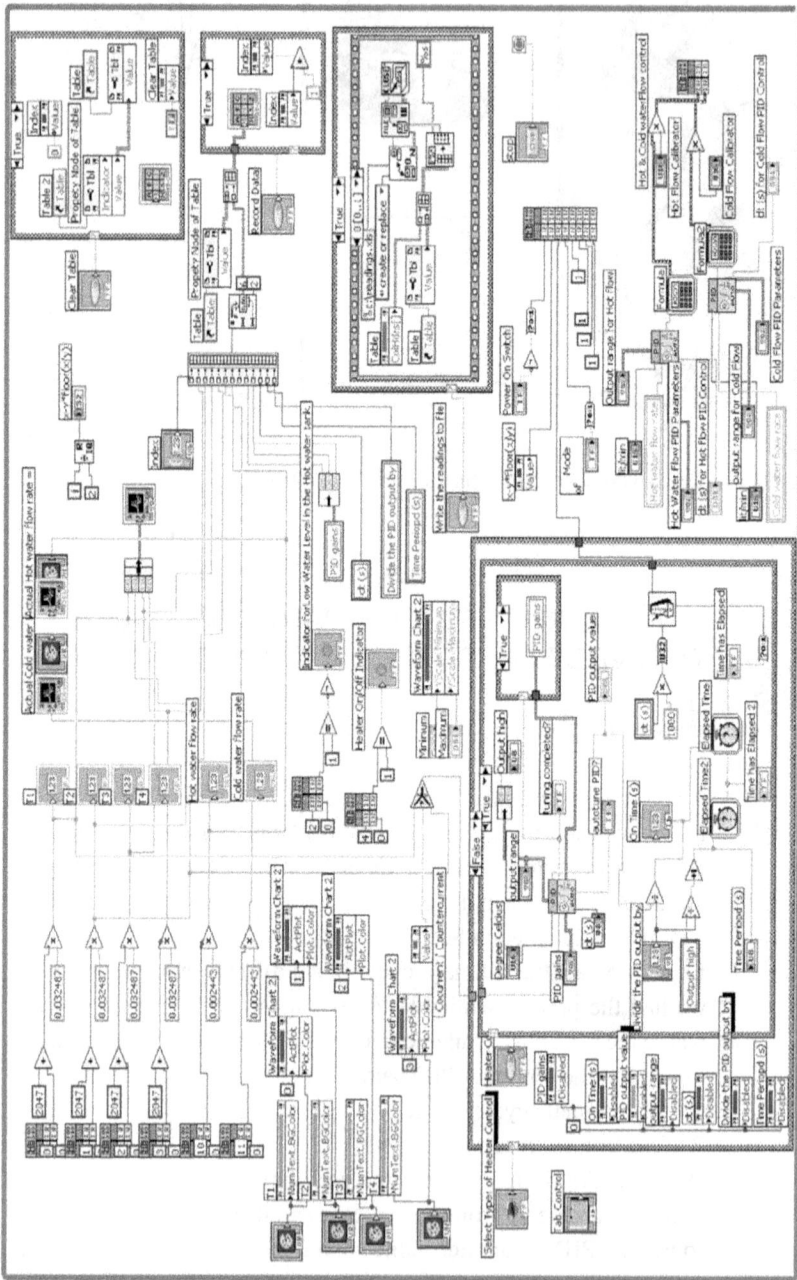

Figure 4.5. Block Diagram of the LabVIEW interface.

4.5.1 ANALYTICAL (MATHEMATICAL DERIVATION)

This section only explains about P control for the tank heater. The rest of the parameters derivation can be done on similar lines.

(Reference: Stephanopoulos, G. 1984. *An Introduction to Theory and Practice*. New York, NY: Prentice Hall.)

Problem Definition: How should Q change in order to keep temp T const and, when Ti changes? (Q, T, and T_i are defined in the following text).

Our system (Armfield's Shell and Tube Heat Exchanger equipment) includes two elements:

1. Continuously stirred tank heater (CSTH)
2. Shell and tube heat exchanger

Both have their own mathematical models. However, we can ignore the heat exchanger model for the time being and only concentrate on the CSTH (Figure 4.6).

Let,

F_i: Flow rate of the water entering the CSTH.

T_i: Temperature of the water entering the CSTH.

F: Flow rate of the water leaving the CSTH.

T: Temperature of the water leaving the CSTH.

Q: Heat energy supplied to the water in the tank by the steam. (Note: In our case the heat energy will be supplied by the electrical heater.)

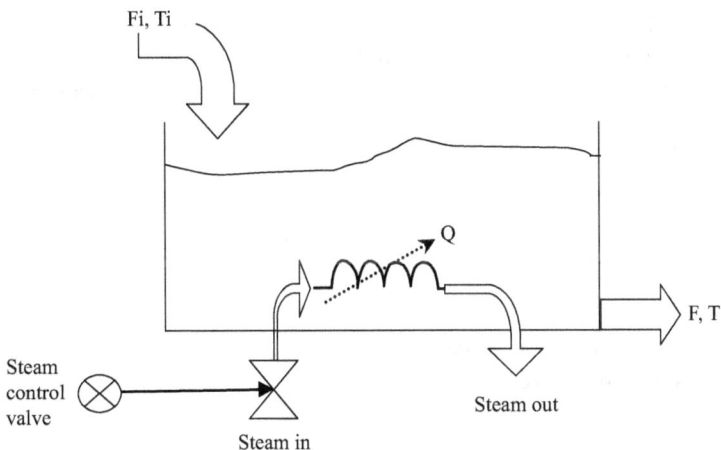

Figure 4.6. Schematic of the CSTH.

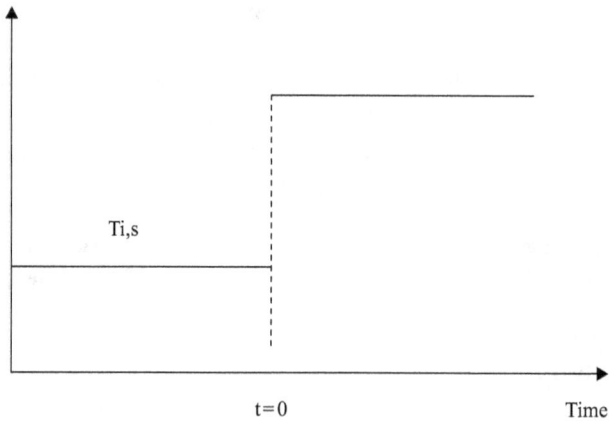

Figure 4.7. Sudden increase in T_i.

Answer:-

In steady state (i.e. T=Ts, V=v) where T is temperature and V is volume of the water in the tank.

The energy balance around the tank yields.

$$0 = F \cdot \xi \cdot Cp \cdot (T_i,s - Ts) + Qs \tag{4.1}$$

where

ξ = Heat coefficient

Cp = Specific heat of water

Now,

Suppose T_i increases suddenly as shown in Figure 4.7.

If nothing is done on Q, the temperature T will start rising with time.

And, h T changes with time will be given by the transient energy balance around the tank.

That is,

$$V \cdot \xi \cdot Cp \cdot \frac{dT}{dt} = F \cdot \xi \cdot Cp \cdot (T_i - T) + Q \tag{4.2}$$

where,

V = volume of the water in the tank.

(4.2) – (4.1) gives

$$V \cdot \xi \cdot Cp \cdot \frac{d(T - Ts)}{dt} = F \cdot \xi \cdot Cp \cdot [(T_i - T_i, s) - (T - Ts)] + [Q - Qs]$$

Accumulation $\in$ MV

$\in = T-Ts =$ Error or deviation of liquid's temp "T" from desired value Ts.

We want to drive "$\in$" to zero by manipulating appropriately the value of heat i/p "Q."

To do this, there are various control laws. We are only interested in PID laws, namely, P, PI, and PID.

For "P" control:

$Q = \alpha(T - Ts)$

$\quad = \alpha \cdot \in$

Put this in Equation 4.2,

$$V \cdot \xi \cdot Cp \cdot \frac{d\in}{dt} = F \cdot \xi \cdot Cp \cdot [(T_i - T_i,s) - \xi] + [\alpha \cdot \xi - Qs]$$

$$= F \cdot \xi \cdot Cp \cdot [x - \in] + [\alpha \cdot \xi - Qs] \dots (x \text{ is constant})$$

$$\frac{d\in}{dt} = F \cdot \xi \cdot Cp \cdot \frac{1}{V \cdot \xi \cdot Cp} \, x - F \cdot \xi \cdot Cp \, \frac{1}{V \cdot \xi \cdot Cp} \in \cdot$$

$$+ [a \cdot \in \cdot \frac{1}{V \cdot \xi \cdot Cp} - Qs \cdot \frac{1}{V \cdot \xi \cdot Cp}]$$

assume $a \cdot \dfrac{1}{V \cdot \xi \cdot Cp} = y$

$Qs \cdot \dfrac{1}{V \cdot \xi \cdot Cp} = z$

y, z are constants.

$$\frac{d\in}{dt} = (\frac{F}{V} \cdot x) - (\frac{F}{V} \cdot \in) + (y \in - z)$$

$$\frac{d\in}{dt} = (\frac{F}{V} \cdot x - z) - \in (\frac{F}{V} + y)$$

$$\underset{a}{\underbrace{\qquad\qquad}} \quad \underset{b}{\underbrace{\qquad\qquad}}$$

$$\int_{\in}^{0} \frac{d\in}{dt} = \int_{0}^{t} dt$$

$$\frac{1}{b} \ln(a + \varepsilon b)_{\in}^{0} = t$$

$$\ln(\frac{a}{a + \in b}) = bt$$

$$\frac{1}{1 + \in (\dfrac{b}{a})} = e^{-bt}$$

$$\in = (a/b)(e^{-bt} - 1)$$

Notice that a and b are constants and b is proportional to α, which is our controlling factor. As α increases, the response is faster.

Thus, by plotting responses of error versus various values of α, we can choose the appropriate value of α.

Similarly, in case of PI control law, is proportional to error and proportional to the time integral of $(T-Ts)$.

According to this P-I control law, the value of the i/p 'Q' is given by:

$$Q = a\,(T - Ts) + \beta \int_0^t (T - Ts)\, dt + Qs.$$

$$Q = a + \beta \int_0^t \in dt + Qs.$$

We can derive similar expression of error as a function of β.

4.5.2 EXPERIMENTAL (ULTIMATE CYCLE METHOD)

1. In this method, the two measurements to be obtained are
   a. Ultimate gain (Gu)
   b. Ultimate period (Pu)
2. The following method is used to find out the preceding two values.
3. First set parameters D=0 and I=0. Then start up the process with P to a low value (say 1).
4. Increase P such that the controlled variable starts to oscillate. (We call it sustained oscillations.)
5. The last value of P that gave the sustained oscillations is the value of Gu. And the period of the oscillations is Pu (measured in minutes).

With these values in hand, we can calculate the PID values from the following formulae.

Case 1) For P only control:
P = 0.5 Gu
Case 2) For PI control:
P = 0.45 Gu
I = 1.2/Pu (min^{-1})
Case 3) For PID control:
P = 0.6 Gu
I = 2.0 / Pu (min^{-1})
D = Pu/8 (min)
1. Calculations and tuning:
We obtained
Pu = 80 s = 1.33 Min
Gu = 282

As we wanted to use PID control, we get the PID values from the formulae as follows:

P = 169.8
I = 1.5
D = 0.166

These values gave me the initial PID values to start the tuning with.

2. Now after putting these values and observing the system response for approximately 5 min, we observed that the amplitude of the oscillations was reduced but the system was moving toward instability (so-called unbounded output).

That means, the integral component was more. We have to reduce "I" by 50%.

Now I = 0.75
P = 169.8
D = 0.166

3. This brought the system back to stability. However, there was a steady state error of almost 1 degree. The parameter that contributes to SS error is P. That means P has to be reduced. Hence "P" was reduced by 50%.

Now P = 85
I = 0.75
D = 0.166

4. There still remained an SS error of 0.5 degree. Hence we reduced P by 50 % again.
Now P = 40 (Instead of 42), I = 0.75, D = 0.166
Also we thought of making it faster. Hence, we doubled the D value to 0.322 and then to 0.4.

4.6   QUESTIONS

1. **What is watchdog circuit? What is the use of it?**
Watchdog circuit is fitted on the HT30XC service unit as a safety measure. This circuit is polled by the computer every 10 s by sending a pulse. If the pulse is not received from the computer in 10 s, this circuit detects that and considers a communication channel failure or a software crash. In this case, this circuit automatically switches off the heater and the pump to avoid overheating and a permanent damage to the service unit.

2. **How is while loop structure utilized in the experiment?**
   The "While Loop structure" accommodates the whole experiment. The "Loop Condition" is tied to a push button. This button is labeled "Start/Stop" and is used for starting and stopping the experiment.

3. **How is the stacked sequence structure used in the VI?**
   The requirement of our experiment is the dependency of the following events in that order:
   1. Start taking the readings.
   2. Save the readings to a file.
   3. Stop taking the readings.

4. **Upload the file to a FTP server immediately.**
   As explained earlier, we want to save the readings of an experiment to a file only after all the readings are available. Once the file is saved to a disk, only then do we want to upload the file to say an FTP server. This is where we have used the stacked sequence structure.

5. **How is the case structure used in the VI?**
   We have two modes of operation for the Heat Exchanger setup, namely, co-current and concurrent. The case structure is used for allowing the user to choose between the two modes of operations. There are two ways to control the temperature, namely, PID control and comparator control. To allow this choice, the case structure is used.

6. **Describe the ultimate cycle method.**
   In 1942, J.G. Ziegler and N.B. Nichols published two methods for tuning P, PI, and PID controllers. These two methods are the Ziegler–Nichols' closed loop method (ultimate cycle method) and the Ziegler–Nichols' open loop method (process reaction-curve method).

7. **What is ultimate gain and ultimate period?**
   Ultimate gain is the maximum value of |Gu| that results in a stable closed-loop system when the proportional-only control is used. Ultimate period is given by the equation $P = \dfrac{2\pi}{\omega_c}$, where ω_c is the value of ω and open loop phase angle is $-180°$.

CHAPTER 5

Experiments

This chapter shows the four experiments that can be performed using the Virtual Lab setup.

Experiment 1	Shell and tube heat exchanger
Experiment 2	Energy balance across a shell and tube heat exchanger
Experiment 3	Temperature efficiencies and temperature profiles of counter-current and co-current mode of operation
Experiment 4	PID control of the heater

5.1 HOW TO PERFORM AN EXPERIMENT USING THE LABVIEW INTERFACE?

The LabVIEW interface is developed such that the user can easily control all the process parameters from a single screen. A separate screen has been developed to take the readings conveniently. There are two experiments that are designed to be performed with Armfield's Heat Exchanger equipment using the LabVIEW interface so far.

1. Heat exchanger experiment
2. PID control of the heater

Instructions to perform these experiments can be found in the appendix at the end of the book.

5.2 HOW WOULD A STUDENT ACCESS THE EXPERIMENT OVER THE INTERNET?

A website is developed, which has a "Perform Experiment" section that can be used to access the experiment over the Internet.

5.2.1 WEBSITE

The website has currently five sections, namely:

1. Home
2. Equipment
3. Instructions
4. Perform experiment
5. Assignments

Home: This section currently has a snapshot of the actual equipment used for the experiment. However, this section gives introductory information about the website.

Equipment: This section gives information about hardware peripherals. It mentions various details about HT30XC Service Unit and HT33 Shell and Tube Heat Exchanger. It also shows schematic as well as pictures of the real equipment.

Instructions: This section has all the instructions required for a user to get his/her computer up for accessing the experiment online. Also, the general instructions required for running the LabVIEW VI are included in this section.

Perform experiment: This section has the actual LabVIEW interface for the experiment. In order to view this interface, one needs to follow the instructions in the instructions section to install all the prerequisite software.

Assignments: This section has experiment-specific assignment files. One can easily download these MSWord files and use them to perform the experiment along with the general instructions provided in the *Instructions* section.

5.2.2 INSTALLING PREREQUISITE SOFTWARE

Two prerequisite software need to be installed on the computer in order to access the experiment online. Following are the instructions to install those prerequisites.

To perform an online experiment you will need,

1. LabVIEW Runtime engine
2. JAVA Runtime engine

You may find both these software on the Internet. Or to install the above-mentioned software, please follow the instructions step by step:

1. Either use your own computer or use those machines that will let you install the above-mentioned software. (*Note*: To install these software, you need to login with Admin rights.)
2. Go to the following link: http://<your_hostname>/Instructions.html
3. Under the heading "Computer Requirements," you will see links to download the above-mentioned software.
4. Download and install both software on your machine.
5. Restart the computer.

When the prerequisite software are installed, you will need to follow the instructions given in the following text before you can actually start with the experiment:

1. Go to the URL http:// <your_hostname>//index.html
2. Click on the "Perform Experiment" hyperlink. It will open a new window that will have the LabVIEW interface for the experiment.
3. Right click on the VI and select "Request Control." You should see a message "Control Granted" in order to proceed with the experiment.

5.3   EXPERIMENT RESULTS

This section shows the sample run of both experiments. As mentioned earlier, there are two experiments designed with the Armfield's heat exchanger and LabVIEW interface. This section shows the sample run of both experiments.

5.3.1   EXPERIMENT 1: SHELL AND TUBE HEAT EXCHANGER

This experiment demonstrates indirect heating or cooling by transfer of heat from one fluid stream to another when separated by a solid wall (fluid to fluid heat transfer).

The input parameters are as follows:

T1 (Temp of hot water in)   : 60°C
Hot water flow rate   : 1 L/min
Cold water flow rate   : 3 L/min

Experiment results using Armfield's interface

Table 0-1: Readings of countercurrent mode of operation

Sample number	Temp T1 (°C)	Temp T2 (°C)	Temp T3 (°C)	Temp T4 (°C)	Hot water pump setting (%)	Hot water flowrate Fhot (L/min)	Cold water valve setting (%)	Cold water flowrate Fcold (L/min)	Flow orientation
1	61.2	49.9	15.5	19.4	19.000	1.02	76	3.02	Countercurrent
2	60.9	49.6	15.5	19.2	19.000	1.00	76	3.00	Countercurrent
3	60.8	49.5	15.5	19.2	19.000	1.00	76	3.00	Countercurrent
4	60.7	49.5	15.5	19.2	19.000	1.04	76	3.00	Countercurrent
5	60.9	49.4	15.5	19.2	19.000	1.02	76	3.02	Countercurrent
6	60.9	49.4	15.5	19.2	19.000	1.01	76	3.03	Countercurrent
7	61.1	49.5	15.5	19.2	19.000	0.95	76	3.00	Countercurrent
8	61.1	49.3	15.5	19.1	19.000	0.98	76	3.02	Countercurrent
9	61.2	49.3	15.5	19.2	19.000	0.99	76	3.01	Countercurrent
10	61.3	49.5	15.5	19.2	19.000	1.01	76	3.00	Countercurrent
11	61.2	49.4	15.5	19.1	19.000	0.99	76	3.03	Countercurrent
12	61.2	49.7	15.5	19.1	19.000	0.99	76	3.04	Countercurrent
13	61.0	49.7	15.5	19.2	19.000	0.98	76	3.02	Countercurrent
14	61.0	49.3	15.5	19.2	19.000	0.96	76	2.99	Countercurrent
15	60.9	49.2	15.5	19.1	19.000	0.98	76	3.00	Countercurrent
16	60.8	49.4	15.5	19.2	19.000	1.00	76	3.02	Countercurrent
17	60.8	48.9	15.4	19.1	19.000	0.97	76	3.02	Countercurrent
18	60.8	49.3	15.4	19.0	19.000	0.96	76	3.07	Countercurrent
19	60.7	48.9	15.4	19.0	19.000	0.98	76	3.06	Countercurrent

Readings for co-current mode of operation

Sample number	Temp T1 (°C)	Temp T2 (°C)	Temp T3 (°C)	Temp T4 (°C)	Hot water pump setting (%)	Hot water flowrate Fhot (L/min)	Cold water valve setting (%)	Cold water flowrate Fcold (L/min)	Flow orientation
1	55.0	62.3	15.4	22.8	33.000	2.01	54	2.08	Co-current
2	55.5	63.1	15.4	22.9	33.000	2.01	54	2.12	Co-current
3	55.4	62.6	15.4	22.9	33.000	2.02	54	2.11	Co-current
4	54.9	62.0	15.4	22.9	33.000	2.04	54	2.11	Co-current
5	54.9	62.1	15.4	22.7	33.000	2.00	54	2.09	Co-current
6	55.0	61.9	15.4	22.8	33.000	2.02	54	2.08	Co-current
7	55.0	62.4	15.4	22.8	33.000	1.99	54	1.92	Co-current
8	55.2	62.4	15.5	23.1	33.000	2.04	54	1.98	Co-current
9	55.7	62.6	15.4	23.2	33.000	2.02	54	2.00	Co-current
10	55.2	61.9	15.4	23.1	33.000	2.01	54	2.00	Co-current
11	55.6	62.9	15.4	23.2	33.000	2.02	54	2.03	Co-current
12	55.0	62.2	15.4	23.0	33.000	2.00	54	2.01	Co-current
13	55.2	62.1	15.4	23.1	33.000	2.05	54	2.00	Co-current
14	55.7	62.9	15.4	23.2	33.000	2.02	54	2.03	Co-current
15	54.9	62.0	15.4	23.2	33.000	2.04	54	1.99	Co-current
16	55.2	62.2	15.4	23.1	33.000	1.99	54	2.00	Co-current
17	55.3	62.4	15.4	23.0	33.000	2.00	54	2.00	Co-current
18	54.9	62.0	15.4	23.1	33.000	2.01	54	1.98	Co-current
19	56.2	63.4	15.4	23.3	33.000	2.03	54	2.01	Co-current

Calculations:

Average values of the quantities:

Mode	T1(°C)	T2(°C)	T3(°C)	T4(°C)	Cold flow (lit/min)	Hot flow (lit/min)
Countercurrent	60.97	49.4	15.4	19.16	0.99	3.01
Co-current	55.25	62.39	15.4	23.0	2.01	2.02

DThot and DTcold in both the modes:

Modes ->	Countercurrent	Co-current
DThot ->	T1 – T2 = 11.56	T2 – T1 =7.13
DTcold ->	T4 – T3 = 3.68	T4 – T3 = 7.615

Result:

This experiment demonstrates that using a simple heat exchanger, a stream of cold fluid can be heated by indirect contact with another fluid stream at a higher temperature (the fluid streams being separated by a wall that conducts heat). This transfer of heat results in a cooling of the hot fluid.

5.3.2 EXPERIMENT 2: ENERGY BALANCE ACROSS A SHELL AND TUBE HEAT EXCHANGER

5.3.2.1 Objective

To perform an energy balance across a shell and tube heat exchanger and to calculate the overall efficiency at different fluid flow rates.

5.3.2.2 Method

By measuring the changes in temperature of the two separate fluid streams in a shell and tube heat exchanger and calculating the heat energy transferred to/from each stream to determine the overall efficiency.

5.3.2.3 Equipment Required

HT30XC Heat Exchanger Service Unit
   HT33 Shell and Tube Heat Exchanger

5.3.2.4 Procedure

5.3.2.4.1 Operational Procedures

1. Enter the temperature controller screen and set the set point to 60°C and mode to automatic.
2. Adjust the cold water flow control (not the pressure regulator) to give 1 L/min and the hot water flow control to give 3 L/min.
3. Allow the heat exchanger to stabilize (use the IFD Channel History screen to monitor the temperatures).
4. When the temperatures are stable, take a sample
5. Adjust the cold water flow control to give 2 L/min. Allow the heat exchanger to stabilize then take another sample.

5.3.2.5 Results and Calculations

Each set of readings is presented in the table.
The columns we are interested in are as follows:

Mass flow rate (hot fluid)	qmh	(kg/s)
Mass flow rate (cold fluid)	qmc	(kg/s)
Heat power emitted	Qe	(W)
Heat power absorbed	Qa	(W)
Heat power lost	Qf	(W)
Overall efficiency	h	(%)

You should also estimate and record the experimental errors for these measurements. Estimate the cumulative influence of the experimental errors on your calculated values for Qe, Qa, Qf, and h.

Compare the heat power emitted from/absorbed by the two fluid streams at the different flow rates.

Heat power:

(Q) = Mass flow rate (qm) × specific heat (Cp) × change in temperature (DT) (W)

Therefore, *heat power emitted from hot fluid:*

$Qe = qmhot \cdot Cphot(T1\text{-}T2)$

$qmhot = 0.017$

$Cp = 4.182$

$DT = 11.56$

$Qe = 0.017 \times 4.182 \times 11.56 = 0.82184664$

Heat power absorbed by cold fluid:

$Qa = qmcold \cdot Cpcold(T4-T3)$
qmcold = 0.050
Cpcold = 4.182
DTcold = 3.68
$Qa = 0.050 \times 4.182 \times 3.68 = 0.769488$
Heat power lost (or gained):
$Qf = Qe - Qa$ (W)
$Qf = 0.05235864$
Overall efficiency:
h = (Qa/Qe)100
= 93.63

5.3.2.6 Conclusions

Theoretically Qe and Qa should be equal. In practice, these differ due to heat losses or gains to/from the environment. In our case Qe value is more than that of Qa. This is because of the heat losses to the environment. As the cold fluid flow rate increases, the rate of heat energy lost to the cold water will increase.

5.3.3 EXPERIMENT 3: TEMPERATURE EFFICIENCIES AND TEMPERATURE PROFILES OF COUNTERCURRENT AND CO-CURRENT MODE OF OPERATION

5.3.3.1 Objective

To demonstrate the differences between co-current flow (flows in same direction) and countercurrent flow (flows in the opposite direction) and the effect on the heat transferred, temperature efficiencies, and temperature profiles through a shell and tube heat exchanger.

5.3.3.2 Method

By measuring the temperatures of the two fluid streams and using the temperature changes and differences to calculate the heat energy transferred and the temperature efficiencies.

5.3.3.3 Equipment Required

HT30XC Heat Exchanger Service Unit
    HT33 Shell and Tube Heat Exchanger

5.3.3.4 Theory/Background

5.3.3.4.1 Countercurrent Operation

Figure 5.1 shows the concurrent operation flow directions of hot and cold water. When the heat exchanger is connected for countercurrent operation, the hot and cold fluid streams flow in opposite directions across the heat transfer surface (the two fluid streams enter the heat exchanger at opposite ends). The hot fluid passes through the seven tubes in parallel, the cold fluid passes across the tubes three times, directed by the baffles inside the shell.

5.3.3.4.2 Co-current Operation

When the heat exchanger is connected for co-current operation, the hot and cold fluid streams flow in the same direction across the heat transfer surface (the two fluid streams enter the heat exchanger at the same end).

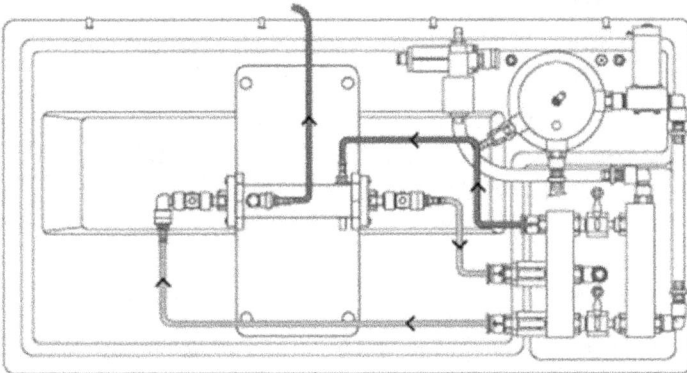

Figure 5.1. Concurrent operation flow directions.

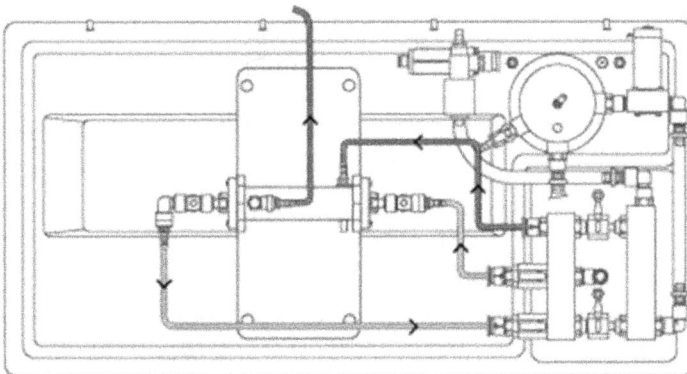

Figure 5.2. Co-current operation flow directions.

5.3.3.5 *Procedure*

5.3.3.5.1 Operational Procedures

1. Enter the temperature controller screen and set the set point to 60°C and mode to automatic.
2. Adjust the cold water flow control (not the pressure regulator) to give 1 L/min and the hot water flow control to give 2 L/min.
3. Allow the heat exchanger to stabilize (use the IFD Channel History screen to monitor the temperatures).
4. When the temperatures are stable, take a sample.
5. Repeat the experiment using co-current flow.

5.3.3.6 *Results and Calculations*

Each set of readings is presented in the table.

Note: In co-current flow T3 is the cold fluid outlet temperature and T4 is the cold fluid inlet temperature.

The columns we are interested in are as follows:

Reduction in hot fluid temperature	DThot (°C)
Increase in cold fluid temperature	DTcold (°C)
Heat power emitted from hot fluid	Qe (W)
Temperature efficiency for hot fluid	Hh(%)
Temperature efficiency for cold fluid	Hc (%)
Mean temperature efficiency	Hm (%)

You should also estimate and record the experimental errors for these measurements. Estimate the cumulative influence of the experimental errors on your calculated values for each of the preceding temperature differences and efficiencies.

5.3.3.6.1 Countercurrent Temperature Profile

From the previous exercises, *the reduction in hot fluid temperature:*

$DThot \quad = T1 - T2 = 11.6$
Increase in cold fluid temperature:
$DTcold \quad = T4 - T3 = 3.7$
Heat power emitted from hot fluid:
$Qe = qmhot \cdot Cphot \cdot DThot$
$Qe = 0.017 \times 4.182 \times 11.6$
$Qe = 0.8246904$

A useful measure of the heat exchanger performance is the temperature efficiency of each fluid stream. The temperature change in each fluid stream is compared with the maximum temperature difference between the two fluid streams giving a comparison with an exchanger of infinite size.

Temperature efficiency for hot fluid:
$Hh = (T1–T2/T1–T3) \cdot 100$
$Hh = (11.6/45.5) \times 100 = 25.49(\%)$
Temperature efficiency for cold fluid:
$Hc = (T4–T3/T1–T3) \cdot 100$
$Hc = (3.7/45.5) \times 100 = 8.13\ (\%)$
Mean temperature efficiency:
$Hm = (Hh+Hc)\ /2$
$\quad\ = (25.49+8.13)/2 = 16.81\ (\%)$

5.3.3.6.2 Co-current Temperature Profile

From the previous exercises, *the reduction in hot fluid temperature:*

$DThot = T1 - T2 = -7.1$
Increase in cold fluid temperature:
$DTcold = T3 - T4 = -7.6$
Heat power emitted from hot fluid:
$Qe = qmhot \cdot Cphot \cdot |\,DThot\,|$
$\quad\ = 0.033 \times 4.184 \times 7.2$
$\quad\ = 0.9941184$
Temperature efficiency for hot fluid:
$Hh = (T1–T2/T1–T4) \cdot 100$
$\quad\ = (7.2/32.2\ ) \times 100$
$\quad\ = 22.36\ (\%)$
Temperature efficiency for cold fluid:
$Hc = (T3–T4/T1–T4) \cdot 100$
$\quad\ = (7.615/32.2) \times 100$
$\quad\ = 23.64\ (\%)$
Mean temperature efficiency:
$Hm = (Hh+Hc)/2$
$\quad\ = 23(\%)$

5.3.3.7 Conclusions

From the mean temperature efficiencies calculated in both cases, it looks like the co-current mode was better than the countercurrent mode of operation in this particular case. However, the selection of the best arrangement for a particular application depends on many parameters such as overall heat transfer coefficient, logarithmic mean temperature difference, fluid flow rate, and so forth. These will be explained and investigated in later exercises.

	DThot	DTcold
Countercurrent	11.6	3.7
Co-current	7.1	7.6

As can be seen from the preceding table, DThot in the case of the countercurrent mode of operation is more than that of the co-current mode. However, the DTcold value is less in case of the countercurrent as compared to its value in case of the co-current mode.

As can be seen from the preceding calculations, in the case of countercurrent mode of operation, hot fluid temperature efficiency is better than that of cold fluid temperature efficiency. But in the case of the co-current mode, both are almost equal.

5.3.4 EXPERIMENT 4: PID CONTROL OF THE HEATER

Aim: To observe the improved system response due to PID control, as compared to the response of the system with simple comparator control.

Two different approaches to this experiment:

1. Approach a: We changed the PID values as per Table 0.3 given in the assignment.
2. Approach b:
   i. We measured the period of the oscillations Pu (in minutes). We knew the value of Gu (current value of P).
   ii. Then using the ultimate cycle method's formulae, we continued tuning the PID parameters until we obtained a smoother system response with an average error = <0.5%. The formulae are as shown in the following text.

Table 0.3. Formulae to calculate the P, I, and D parameters in the experiment

Control Law	Formulae for P	Formulae for I	Formulae for D
P	$0.5 \times$ Gu	–	–
PI	$0.45 \times$ Gu	1.2/Pu	–
PID	$0.6 \times$ Gu	2/Pu	Pu/8

Graphs:
Figure 5.3 provides a graph of the controlled temperature variable ($T1$).

Result:

Control	Average error in steady state
Comparator control	1.2%
PID control	0.17%

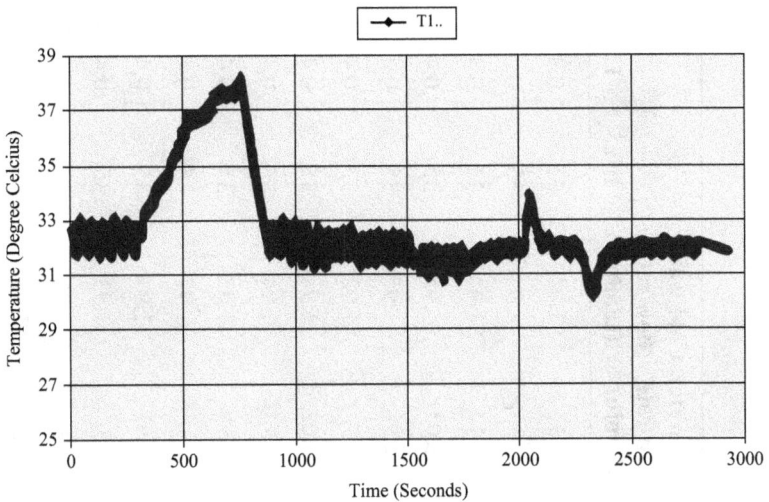

Figure 5.3. System response (temperature [T1] versus time).

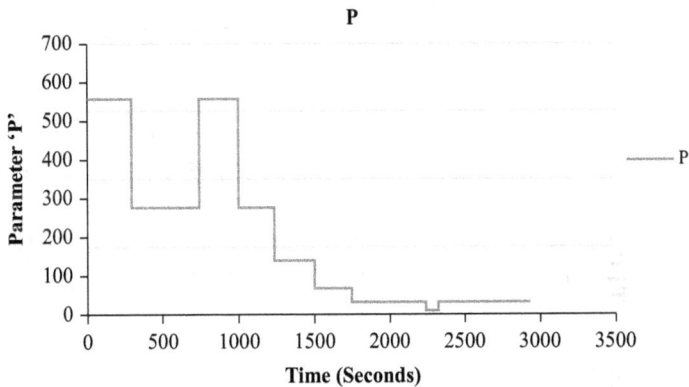

Figure 5.4. Parameter "P" versus time.

Table of Readings

Sample No.	T1 (°C)	T2 (°C)	T3 (°C)	T4 (°C)	Hot water flow rate (L/min)	Cold water flow rate (L/min)	P	I	D	dt (s)	Divide the PID output by	Time period (s)
2599	32.1	29.4	7.86	11.79	3	1.94	35	0.4	0	1	7	15
2600	32.23	29.53	7.86	11.83	3	1.94	35	0.4	0	1	7	15
2601	32.19	29.63	7.86	11.76	3	1.94	35	0.4	0	1	7	15
2602	32.19	29.63	7.8	11.83	3.01	1.95	35	0.4	0	1	7	15
2603	32.19	29.63	7.86	11.86	2.98	1.95	35	0.4	0	1	7	15
2604	32.16	29.66	7.86	11.79	3	1.92	35	0.4	0	1	7	15
2605	32.16	29.63	7.86	11.89	3.02	1.94	35	0.4	0	1	7	15
2606	32.16	29.63	7.96	11.83	3	1.94	35	0.4	0	1	7	15
2607	31.97	29.53	7.86	11.92	2.99	1.92	35	0.4	0	1	7	15
2608	31.93	29.47	7.86	11.86	3	1.97	35	0.4	0	1	7	15
2609	31.97	29.47	7.86	11.89	3.01	1.97	35	0.4	0	1	7	15
2610	31.87	29.43	7.86	11.79	3	1.93	35	0.4	0	1	7	15
2611	31.84	29.34	7.86	11.89	3	1.97	35	0.4	0	1	7	15
2612	31.87	29.34	7.83	11.79	3	1.98	35	0.4	0	1	7	15

2613	31.9	29.37	7.86	11.83	3	1.93	35	0.4	0	1	7	15
2614	31.9	29.34	7.86	11.76	2.99	1.93	35	0.4	0	1	7	15
2615	32.03	29.37	7.86	11.79	3	2.03	35	0.4	0	1	7	15
2616	32.1	29.5	7.86	11.79	3.03	2.03	35	0.4	0	1	7	15
2617	32.16	29.56	7.86	11.79	3.01	2.03	35	0.4	0	1	7	15
2618	32.19	29.6	7.86	11.79	2.98	2.04	35	0.4	0	1	7	15
2619	32.16	29.6	7.83	11.79	2.99	2.05	35	0.4	0	1	7	15
2620	32.13	29.6	7.86	11.7	3.01	2.03	35	0.4	0	1	7	15
2621	32	29.5	7.83	11.73	2.99	2.01	35	0.4	0	1	7	15
2622	32	29.47	7.86	11.76	2.99	2.07	35	0.4	0	1	7	15
2623	31.97	29.47	7.86	11.7	2.99	2.08	35	0.4	0	1	7	15
Average	32.055	29.508	7.858	11.806	2.9996	1.9796	35	0.4	0	1	7	15

Note: It is not feasible to include all the 2932 readings in the report. Hence, we have included only that part of the readings where the system achieved stability and relatively better response (i.e., response with least average error). However you can see all the data points on the graphs.

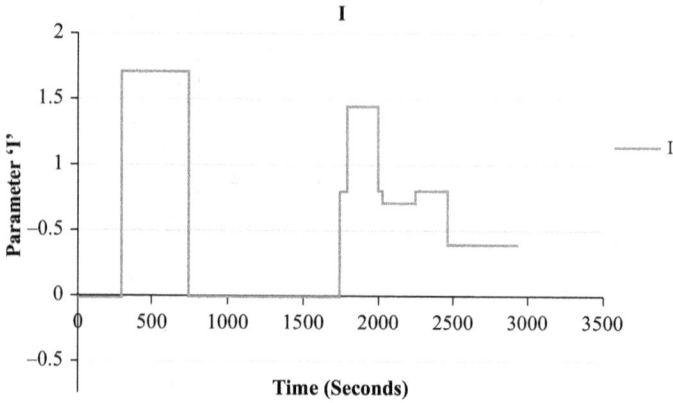

Figure 5.5. Parameter "I" versus time

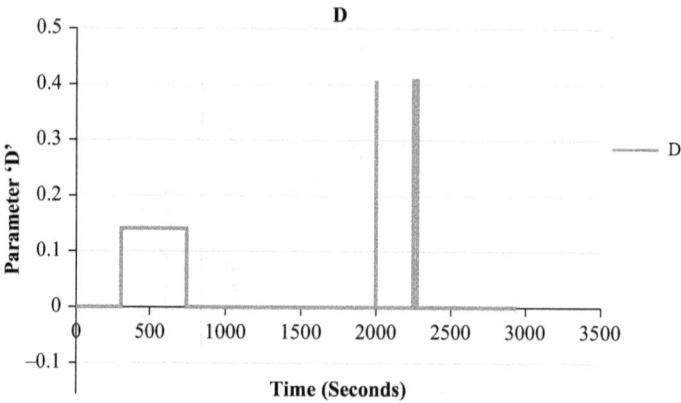

Figure 5.6. Parameter "D" versus time.

From the preceding table, it can be seen that the system response was better due to PID control, as compared to the response of the system with simple comparator control.

5.3.4.1 Feedback from the Students' and Lab Assistant's Observations

This section includes a report submitted by the lab assistant on his own observation of students during the experiment. In the following report "I" refers to the lab assistant.

Report on heat exchanger virtual lab Session 1 held on October 23, 2007, between 8.20 a.m. and 11.00 a.m. at Engineering Technology.

Students who attended virtual lab Session 1:

1. Richard Faber
2. Craig
3. Jason
4. Maggie

I noted the following things as the students performed their experiment in the virtual lab session:

Richard

Richard started at 8.26 a.m. and he was done by 8.52 a.m.

1. Richard read the instructions page, but he tended to follow instructions given in the assignment. So we have to update assignments with the instructions.
2. He was searching for the controls on the VI.
3. I have to decompose the instructions further to make them absolutely stepwise.
4. I have to make students aware about the novel way of taking readings.
5. Also, I have to tell them about the vertical and horizontal scroll bars that get hidden.
6. After taking the readings Richard forgot to release the control. So I will have to make some arrangement to take care of that.
7. It was difficult to find the Download file hyperlink at the bottom of the page, because instructions to which page they should download the reading file from were not clear. Also the link is almost hidden at the bottom of the page.
8. When Richard gained control the second time, suddenly the VI did not show all the things inside. The solution to this problem was that he had to release the control, refresh the page, and request the control once again.

Craig

Craig performed the experiment between 8.55 a.m. and 9.15 a.m.

1. Setting the hot water flow rate can be made easier. I have to think about it.
2. Reduce the three digits of accuracy to two digits.
3. Craig faced trouble setting the inputs due to the i/p o/p difference error. The solution can be reduced from three decimal digits of accuracy to one digit.

Jason

1. He faced trouble accessing the VI because by the time he took over, the wireless Internet signal was too low to access any web page. Hence he used wired Internet to access the lab.
2. During his turn, I noted that nobody had bothered to wait until the temperature reached 60°C before they could start recording the readings.
3. So I had Richard and Jason repeat the experiment and take the readings at temperature $T1 = 60°C$.

Overall conclusion drawn from a chat with Richard, Craig, and Jasonafter the session

1. Richard seemed happy with the so-called toy.
2. Craig said, "It was not a bad experience though." I perceive his statement as "It was OK."
3. Jason faced lot of troubles while doing the experiment. He said that he was not that computer savvy. But I think when he repeated the experiment; he was pretty good with it.
4. I do not know about Maggie. She finished the experiment in a jiffy and left before I could see her perform the experiment.
5. I have to improve the control accuracy for the hot and cold flow rates. Also a control over temperature needs to be improved.

5.3.4.2 Areas of Improvement Identified From the Students' Feedback

1. More accuracy in hot and cold flow controls needed irrespective of any nonlinearity involved.
2. More accuracy and sustainability in the temperature control needed irrespective of the tap water temperature variation.
3. Improvement in the instructions sheet expected.

5.3.4.3 Solutions Implemented for the Above Problems

1. The first two problems were cured by introducing the PID control technique. The initial P, I, and D parameter values were found using "the ultimate cycle" method. Then the parameters were tuned to give least possible error. The method is explained in the "How did I arrive at the PID parameters' values?" section of the report.

2. In the case of a second problem, I talked with every batch about the instruction sheet. I refined the instructions iteratively after every session of the PID lab depending on the students' feedback.

3. I also added a visual aid feature in the LabVIEW interface. This feature lets the student select the color for temperature nodes (T1, T2, T3, T4) on the schematic dynamically during the experiment. The selected color automatically gets applied to the respective graphs of the temperature nodes. This change in the graph's color makes it easier for the student to identify which graph in the temperature chart corresponds to which temperature node.

5.3.5 FEEDBACK FROM STUDENTS ABOUT THE PID EXPERIMENT

This experiment was developed after the suggestions of the students from the first experiment was implemented. This resulted in a better feedback from the students for this experiment.

Feedback from lab assistant: The lab assistant noted that approach "b" is better than approach "a" used for this experiment. Since in approach "a," the automated VI takes care of all the inputs automatically, the student simply sits back and observes the response of the system as the input parameters are changed automatically. This approach rids the student of all the other activities like calculation of P, I, D parameters, thus letting him/her fully concentrate on observing how the system behaves for different P, I, D values. But it was found that students preferred active involvement in the experiment rather than just sitting and watching the system response.

CHAPTER 6

FACTORS INFLUENCING
THE VIRTUAL LAB

This chapter discusses cross-cutting concerns of "Virtual Laboratory," such as drivers, concurrent access, and user authentication.

6.1   DRIVERS FOR PROGRAMMABLE DEVICES

Software drivers for programmable devices are at the lowest level of the Virtual Lab system. Depending on the type of programmable devices, they directly access the hardware or use the manufacturer's software libraries for accessing hardware. They are designed for efficiency and reuse, as every experiment must use one or more programmable devices. Every programmable device is represented by one class or hierarchy of classes. The structure of the device determines the corresponding structure of classes that represent the programming model for each type of programmable device. The top class represents common features of the instrument, while the descendent classes that inherit the top class correspond to logical or physical parts of the device. Setting the devices and functional properties of devices are modeled as class methods, which are accessed by calling corresponding methods of instantiated classes. Programmable devices that are used in the prototype system consist of a data acquisition system (DAS). It has four standard sections:

- Analog digital (A/D) section with 8/16 differential/single ended inputs, 12 bit resolution, with programmable gain amplifier (PGA). A/D is used for direct voltage measurements in experiment circuits and can be also used for indirect measurement of non-electrical quantities by using the appropriate converters with standard voltage output. The measurement of electrical current with A/D is also possible, but by indirectly measuring the voltage on a resistor with known value.

- Digital analog (D/A) section with two channel voltage/current outputs, 12 bit resolution. D/A can be used for the generation of necessary voltages and currents in experiments for two basic purposes. One purpose is to provide necessary conditions for experiment performance. Depending on the particular experiment, voltages/currents can be constant acting as parameters, or changing when it is important to measure and show the functional dependence between observed values, with generated voltage as independent variable. The other purpose is to control some experimental equipment that is not programmable, but that can be managed by using voltage or current signal.

- Digital input/output (DIO) section with four digital 8 bit I/O ports. Each port can be programmed either as input or output. When acting as output, it can be used similar to the D/A section for controlling equipment.

- Programmatic access to DAS is possible using software drivers, and specialized software such as National Instruments LabVIEW, Visual Designer. A direct access from program to registers in memory map is also possible. Registers control the behavior, determine the status of DAS, and contain the measured values. Programming with registers is actually used in the Virtual Lab system.

6.2 CONCURRENT REQUIREMENTS FOR SAME EXPERIMENT

As the Virtual Lab system is designed to be used by many independent users, it is quite possible that two users try to start the same experiment simultaneously, or try to start the experiment while it is already in process—started by some other user. In the case of basic configuration described here, there is physically only one set of programmable devices for each experiment. Depending on the type of programmable device, it may be possible to use the same device simultaneously for different purposes. The word simultaneously means, of course, parallel tasks that are executed on multitasking operating system.

In this implementation, simultaneous access to the same experiment is not allowed, while simultaneous access to different experiments is allowed, provided that they do not use the same hardware components. It is achieved on the side of web server, by using application variables.

Application variables keep the current state of each experiment, and in each user session the value of application variables is examined. Depending on the state of experiment, the user can start the experiment if it is not already started, or get the message that the experiment is in process, and to try later.

6.3 USER AUTHENTICATION

Currently the experiment does not need any login. We do not see any need of the authentication process presently because there are only a handful of students who are accessing the experiment that too only during their lab session, when the lab assistant is available in the lab. In short, there is nothing much on the site other than the LabVIEW VI that is vulnerable to attacks. Here we are assuming that no hacker would like to prove himself by hacking an academic site, which has absolutely no material of general interest, but the heat exchanger experiment specific information. However, in the future, if need arise, the user authentication protocol can easily be incorporated.

6.4 ISSUES SURROUNDING LIVE TRAINING

6.4.1 DEDICATED CLASSES

Besides the general methods and tools for delivering training discussed already, it is important to consider when live training can address the challenges of delivering complex content, and when live training falls short. Dedicated classes, consisting of hands-on instruction taught in physical classrooms, are effective but not scalable to reach a large number of learners. They also are extremely expensive, particularly when limited resources (equipment and instructors) must be sent into the field. Dedicated classes may require any or all of the following:

- Provisioning, configuration, and validation of equipment and software (not to mention the purchase expense)
- Provisioning of trainers
- Travel (trainers and sometimes learners)
- Technical support to ensure "everything works"
- Teardown, packing, and shipping—and perhaps travel on to the next live site destination.

6.4.2 VIRTUAL CLASSROOMS

Virtual classroom is a highly effective and more scalable approach to reach a larger number of students than is live, "bricks and mortar" training. But even virtual classrooms sometimes have their limitations, in terms of best types of training content delivered, instructor student communications, effective automated processes for scheduling, and management of learning objectives. It can be challenging for an instructor using a web conferencing tool to be absolutely certain his students are "getting it," even with polling and testing capabilities commonly found in some of the e-Learning tools. Similarly, a virtual classroom is still a remnant of the model whereby the instructor "does" and the students "watch." Archiving content to make it available for later viewing, while a key aspect of reaching students anytime and anyplace, does not guarantee that the instructor will ever see what the learner learned. Similarly, a web-based lab is fine for reaching a large number of learners, but in no way can it be described as immersive or compelling—nor do they offer sufficient measurements to enable tracking of student progress. They typically rely on a learner learning by observation.

6.4.3 COURSEWARE

Delivery of authored courseware via CDs or DVDs can be most effective at reaching a large number of learners—but like asynchronous libraries, there is little method of ensuring that the learners learn and little ability to adapt the training if conditions change. Using these media can be effective to deliver rich media-based training (such as animations, videos, and simulations) but can be extremely expensive to produce, deliver, and maintain. And they often overlook the value of the live instructor; there's nothing wrong with self-paced materials, but they place the onus for learning on the student without backup of management or measurements. They rely on a student learning by "reading," "watching," or "simulating" the experience.

Lower down the content spectrum is very broadly used: popular tools such as PowerPoint slide shows, Flash demos, and video clips that demonstrate activities. These often are provided as standalone, passive training modules and often provide no means of measurement. This is not to dismiss as invalid the tools under discussion here. But content and courseware are not an effective substitute for context and competency. All too often training organizations spend great efforts at creating the content, and all too often they fail to deliver it in such a way as to guarantee that competency-based learning—which is required to build and maintain software expertise—has taken place.

6.4.4 LEARNING GAP

After considering the tools and methodologies and the business issues surrounding technical training, the final item to consider is the learner himself. Consider the typical cell phone user. He may or may not read the manual(s) that come with the new model cell phone. But he will acknowledge that without actually beginning to use the cell phone, he would never claim to be proficient. One must place calls to learn to use a new phone. And the same applies to virtually any technology in modern life, from microwaves to VCRs. We learn by doing. And nothing is more critical than in the area of technical training, which carries with it so many dependencies and complexities. In many respects the "near hands-on" aspect of simulations comes closest to providing the immersion necessary to learn new technologies. But simulations, while fine for some purposes, lack one thing: a compelling and personalized experience that gets built based on the learner's state of interaction with the learning objects. A newer form of technology, with a valuable underlying pedagogical approach to back it up, has only recently begun to be noticed in e-Learning and training circles. It is still in its infancy but is already being used successfully by early adopters in the field of technical training, and it is called Virtual Labs.

6.5 SYSTEM MANAGEMENT AND DEVELOPMENT

The Virtual Lab system can be useful and used by many categories of users ranging from regular students at various levels of education system, to remote users that relay especially on remote access to education materials. Virtual Lab could be used as an addition to regular laboratory exercises, or if significantly developed as replacement of part or all exercises in particular subjects. It can also be used as part of the educational process not just for remote access to experiments, but as a system to be studied from various aspects—web server programming, software experiment management, management of programmable devices, network connections of PCs that constitute the Virtual Lab system, and so forth. This approach can lead to further system development by advanced students as part of their seminal, graduate, or postgraduate activities, exams or some other kinds of engagement. Finally, this system does not have to be used only for educational purposes, as its structure is general enough to be used in other areas also, such as for monitoring and control of remote systems, or for control and management of some other kind of experiments in real research.

The use of the system can be for educational purposes, but also for some real-life tasks such as monitoring and control of remote systems or for experiments in scientific research. There are two main strategies to increase the number of experiments controlled by Virtual Lab. One is to increase the number of programmable devices I/O for measurement and control, which will enable more experiments to be performed, as the I/O is devoted only to one experiment. The other is to use existing I/O for measurements and control and to introduce switching circuits that will use existing I/O for more than one experiment. That approach also requires increasing the number of I/O for the management of switching circuits. In both approaches, it is necessary to install new PC cards with programmable devices. As each PC can support just a finite number of expansion slots, the introduction of new PCs might be necessary. However, adding new PCs for the control of programmable equipment is very easy in distributed configuration. It is necessary to connect new PCs on the network and to add software for programmable devices and new experiment.

6.6 FUTURE DEVELOPMENTS

- *Experiment management system*: Rather than implementing a separate user authentication process, we can use the University of Toledo's login. We can also have a database of the students that will store the student credentials, his/her experiment results, feedback, and so forth. Thus, we can create a report at the end of the semester to review the overall online experiment experience of students.
- *User authentication*: Currently since the users (students) of the existing system are limited, having an authentication process is not pragmatic, for the sole reason that there is no need of secrecy, nothing much that anybody would want to steal or hack. However, in the future as the number of users increases, the authentication process may become necessary. However, it would be nice to use the existing university logins rather than making the students create a separate login, which they will use only once in a semester.
- *Chatserver*: Chat functionality can be included on the same web page as that of the experiment.
- *Network of virtual labs*: If we make our system robust enough, we

can join other virtual lab projects in other universities to weave a network of virtual labs.

- *Telephone controlled equipment*: As we know, the online lab consists of two things, namely, the "LabVIEW VI" and "the heat exchanger equipment." Of these, the LabVIEW VI is available online 24×7, however, the actual heat exchanger equipment is not,—simply because we do not want to leave the equipment switched on forever. Hence, currently it takes a person to hook up the USB cable to the equipment and turn its switch on in order for the equipment to come to life, before any experiment can be performed. We can obviate the need of a person's presence for this job by making use of a "telephone controlled switch." Then that person can turn the equipment on or off from a distance, whenever a student demands access and schedules a time slot for the experiment.

- *Simulation*: Since allowing a student to play with the actual system in a random manner is not feasible, a simulator can be designed to give a virtual idea to the student about the system. By simulating the real system, the student will get an excellent grasp of the real operation of the system. That way they will have a better idea about what to expect as an output of the experiment that they will perform on the real equipment over the Internet. Thereby, we let the student verify the simulated output with the real-time output, which indeed is an excellent way to reinforce concepts (of PID or) from the course.

CHAPTER 7

Experiment Instructions

This section includes the instructions to be used to perform the experiments mentioned in this document.

7.1 INSTRUCTIONS FOR THE SHELL AND TUBE HEAT EXCHANGER EXPERIMENT

Objective: To demonstrate indirect heating or cooling by transfer of heat from one fluid stream to another when separated by a solid wall (fluid to fluid heat transfer).

Equipment required: HT30XC Heat Exchanger Service Unit, HT33 Shell and Tube Heat Exchanger.

Theory/background: Any temperature difference across the metal plates will result in the transfer of heat between the two fluid streams. The hot water flowing through the inner tube bundle will be cooled and the cold water flowing through the outer shell will be heated.

Note: For this demonstration the heat exchanger is configured with the two streams flowing in opposite directions (countercurrent flow). The cold fluid flowing through the shell is forced to flow over and under baffles in the shell that force the fluid to flow across the tube bundle to improve the heat exchange.

Operational Procedures

1. Enter the temperature controller screen and set the set point to 60°C and mode to automatic.
2. Adjust the cold water flow control (not the pressure regulator) to give 1 L/min and the hot water flow control to give 3 L/min.
3. Allow the heat exchanger to stabilize (use the IFD Channel History screen to monitor the temperatures).

4. When the temperatures are stable take a sample.
5. Adjust the cold water flow control to give 2 L/min. Allow the heat exchanger to stabilize then take another sample.

Results and Calculations

Each set of readings is presented in the table.

The columns we are interested in are the reduction in hot fluid temperature:

$DThot = T2 - T1$ (°C)

And the increase in cold fluid temperature:

$DTcold = T4 - T3$ (°C)

Estimate and record the experimental errors for these measurements. Estimate the cumulative influence of the experimental errors on your calculated values for DThot and DTcold.

Compare the changes in temperature at the different flow rates. If time permits try different combinations of hot and cold fluid flow rate.

Conclusions: You have demonstrated how, using a simple heat exchanger, a stream of cold fluid can be heated by indirect contact with another fluid stream at a higher temperature (the fluid streams being separated by a wall that conducts heat). This transfer of heat results in a cooling of the hot fluid. Comment on the changes in DThot and DTcold when the flow of cold water is increased. The consequence of these changes will be investigated in a later exercise.

7.2   INSTRUCTIONS FOR THE PID CONTROL FOR HEATER EXPERIMENT

Aim: To observe the improved system response due to PID control, as compared to the response of the system with simple comparator control.

Apparatus: Armfield's heat exchanger equipment, LabVIEW control interface, or Armfield's control interface, USB cable.

Instructions:

(The following first four instructions are for the LabVIEW interface user [mostly the lab assistant.])

(*Note*: If you are accessing the experiment over the Internet, simply right click on the VI that appears in the browser after you access the experiment and "Request Control" of the VI. When a message "Control Granted" appears, follow from the fifth instruction given in the following text.)

1. Plug the USB cable in the Armfield's heat exchanger unit and turn on the switch of the unit.

2. Turn the Lab Computer on.
3. Run LabVIEW.
4. Open the VI named "Heatexchanger_with_PID.vi" from the recently opened VIs.
5. Go to the "Pre-lab Instructions" tab on the VI and read the instructions carefully.
6. Go to the "Controls" tab. Observe and understand the schematic picture of heat exchanger shown in here.
7. Run the VI using the Run button. When the VI is run, the "Hot water level indicator" will illuminate.
8. Ensure that the hot water level indicator is green. A red light indicates that the water level is low in the heater tank. Please ask the lab assistant to fill the water in the tank.
9. Before turning on the Power switch on the VI, set the mode of operation as required (Concurrent: hot and cold water flowing in same direction. Countercurrent: hot and cold water flowing in opposite direction).
10. Also select control as "PID control" on the VI.
11. In the PID gains Box, input the following parameters:
    a. *Output high* = 100
    b. *Output low* = 0
    c. *Dt (s)* = 1
    d. *Divide the output by* = 7
12. In the Parameters Input Box, input the following parameters and then notice the corresponding three graphs, shown in the VI, rise up to the value that was set earlier:
    a. *Hot water flow rate: 3 L/min*
    b. *Cold water flow rate: 2 L/min*
    c. *Temperature set point: 32°C*
13. Now, turn on the Power switch on the VI. A green LED on the switch will glow, which indicates that the heat exchanger's internal relay is on and the unit is ready to be used.
14. Now go to the "Readings" tab and hit the "Record Data" button. Record all the readings throughout the experiment. Stop recording the data only at the end of the experiment.
15. Once the temperature reaches almost the set point value, the heater indicator will start going on and off more frequently. The temperature graph will oscillate about the set point value. When the experiment reaches this state, wait for five cycles.
16. Now there are two approaches to this particular step:
    a. *Follow the PID values mentioned in Table 7.1*: Change the PID parameters as given in Table 7.1. After changing the parameters,

observe the response for 1 min and comment on the system's response to the newly entered PID values.

   b. *Use your own values during the experiment*: Use the first four sets of PID values from Table 7.1 to begin with. Then use the ultimate cycle method's formulae to arrive at the optimized PID parameter values. Find the ultimate cycle method explained in any of the Control Systems Theory book (or ask the TA about it during the experiment).

17. Notice that the temperature fluctuations should reduce eventually as the PID parameters are tuned and the response should become smoother over time, with less over or undershoots. This suggests that the steady-state error is reducing. Thus, we have a better control over the temperature of the hot water at the end of the experiment.

18. Note down the observations for each set of PID parameters. Comment on how sensitive the system response was to the "D" parameter in Table 7.1, if following approach "a." Also include Table 7.1 in the report. If following approach "b," prepare a table similar to Table 7.1.

19. Once better control is achieved, measure the relative average percentage error. (*Note: % Error <= 0.5%.*)

20. At the end of the experiment, just to appreciate the better control achieved using PID, switch the mode of control to *Comparator control* and compare the response. Notice that the average error has increased. Measure the average error as percentage of setpoint temperature for this case.

Table 7.1. PID values for approach "a" in step 16

Seq. #	P	I	D
1	560	0	0
2	280	1.71	0.14
3	560	0	0
4	280	0	0
5	140	0	0
6	70	0	0
7	35	1.44	0
8	35	0.72	0
9	17	0.8	0.001

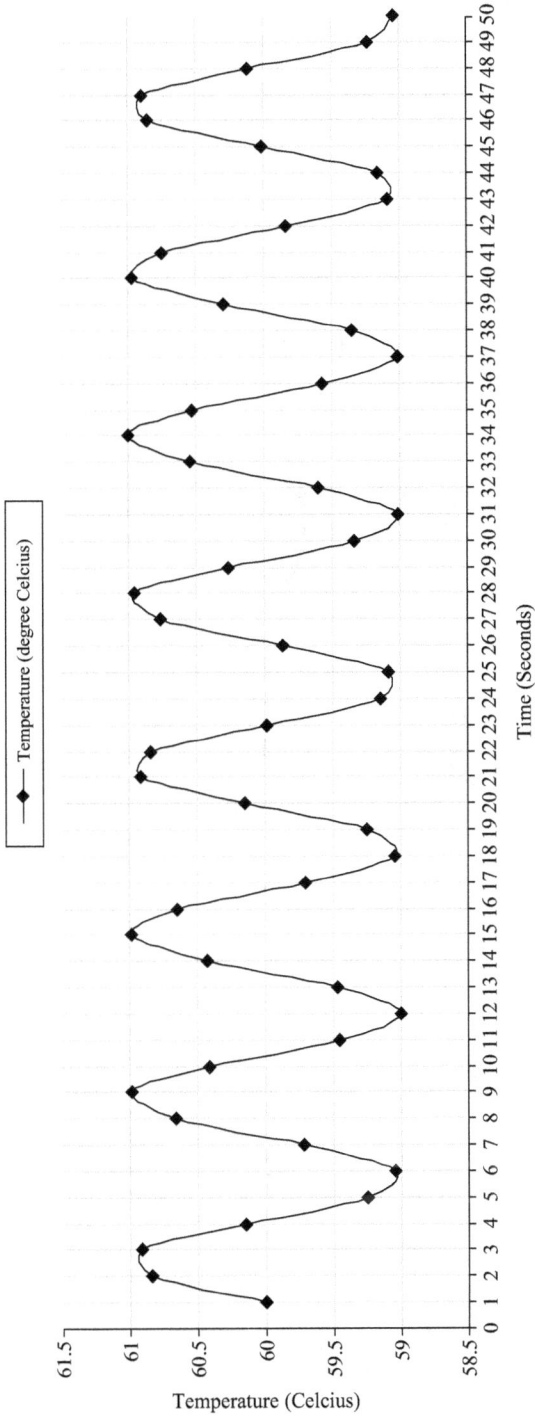

Figure 7.1. Sustained oscillations in the system response.

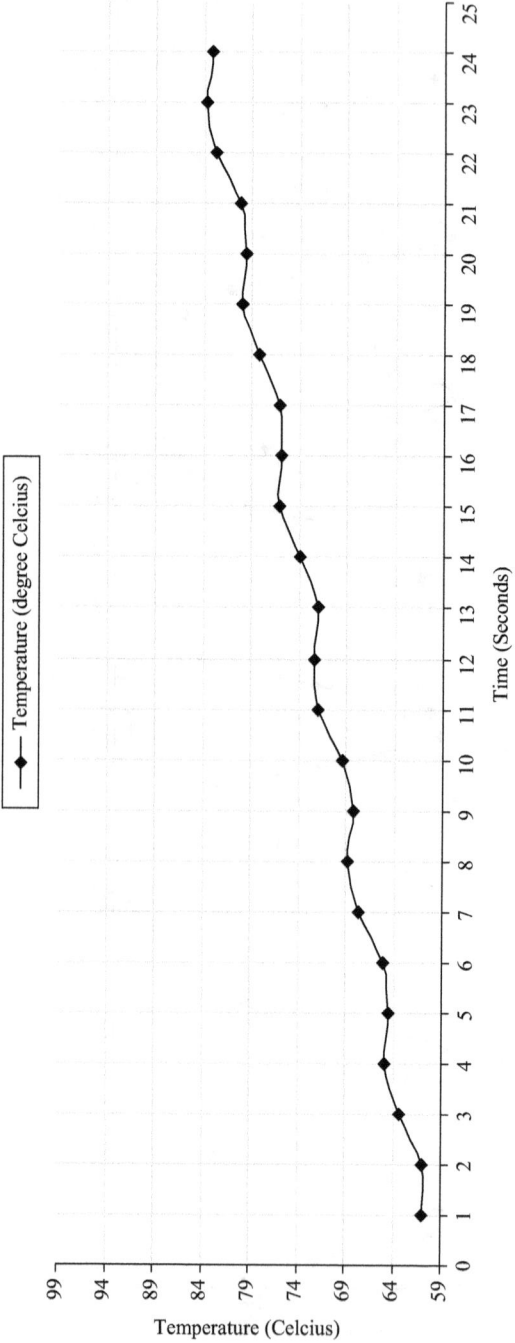

Figure 7.2. Unbounded output of the system.

Please answer the following questions after the PID experiment.

1. What is a "Good Control" from your perspective with respect to the experiment?
2. After using the "comparator control" and "PID control" during the experiment, which one do you think is a better control method? Why do you think so?
3. What is the full form of PID?
4. Explain in your own words how P, I, and D individually contribute to make our system behave better as compared to the system's behavior with a simple comparator control.
5. Solve the following problem using the method you used during the experiment.

You are assigned a task of finding out the PID parameters for the newly brought heat exchanger system in the lab. You decide to use the ultimate cycle method. Let us say for the initial PID values of $P = 50$, $I = 0$, $D = 0$, the heat exchanger is behaving as shown in Figure 7.1. The temperature setpoint is 60°C. What will you change and by how much as the first step toward the tuning of PID parameters?

After you made the change in the PID parameters, you got the following response (Figure 7.2). Is the system stable? Why? Which parameters would you change and by how much such that the system regains its stability? Give reasons for your decision.

CHAPTER 8

RELATED WORK

There are a small number of past and ongoing activities to create and establish online lab facilities. This book presents a new model for interactive virtual laboratory experiments based on LabVIEW. Based on this model, we designed a generic software framework for laboratory instruction over the Internet. We implemented and deployed this framework for a thermodynamics lab. Our results show that this framework can enable more students to be exposed to a comprehensive laboratory experience and increases the involvement of the teaching staff in laboratory instruction.

Dr. Clark K. Colton, a professor from the department of Chemical Engineering, Massachusetts Institute of Technology, coined the term iLabs, referring to their interdepartmental online lab collaborations using VPN. Another prominent example is EGEE, providing a grid infrastructure to distribute and process the vast amount of data resulting from experiments in the large hadron collider at CERN. The U.K. e-Science Program features a working group ("Instruments on the Grid") addressing, for example, the integration of X-ray crystallography or sensors for urban pollution into the grid. During iGrid 2005, the transparent operation of a biology experiment on a test-bed of globally distributed visualization, storage, computational, and network resources was demonstrated. The environment was based on the distributed virtual computer. However, these approaches have a number of different drawbacks: They tend to be either too focused with respect to the targeted instruments, or do not address interactive steering of instruments, or are part of a dedicated demo environment where the environment set-up making the laboratory equipment available is mostly performed manually by the participating researchers. A more generic approach, driven by an initiative supported by the U.S. National Science Foundation, led to the definition of the Common Instrument Middleware Architecture (CIMA). CIMA is targeting on developing a web-service-based middleware stack allowing treating arbitrary instruments as grid resources. Current related projects are X-ray crystallography and the automated observatory. CIMA could become a generic interface to instruments and we will evaluate this stack once it becomes available.

BIBLIOGRAPHY

BOOKS:

Travis, J., and J. Kring. *LabVIEW For Everyone*. Stephanopoulos, G. 1984. *Chemical Process Control: An Introduction to Theory and Practice*. New York, NY: Prentice Hall.

WEBSITES:

NI Developer zone at http://www.ni.com
http://heatex.mit.edu/

ABOUT THE AUTHORS

Ella Fridman, PhD was assistant professor and graduate program director at University of Toledo until 2008 when she died in a car accident. She started as junior scientist at the Institute of Thermophysics of the Ukrainian Academy of Science. In 1989 Ella Fridman became a PhD in Mechanical Engineering and continued to work in the same Institute as senior scientists up to 1990 when she immigrated to the United States.

Harshad Mahajan, MS, graduated from University of Toledo. He worked as a software development specialist at Dox Systems LLC for three years before moving back to India. In India, he did freelancing and online tutoring. He has also worked with esteemed organizations like J.P. Morgan Chase and L&T Infotech. He presently works as a Java application developer in Mumbai and can be reached via email at mahajan.harshad@gmail.com.

INDEX

FORTHCOMING TITLES FROM OUR THERMAL SCIENCE AND ENERGY COLLECTION

Derek Dunn-Rankin, Editor

Advanced Technologies in Biodiesel: Introduction to Principles and Emerging Trends 12/31/2014
By Aminul Islam

Advanced Technologies in Biodiesel: New Advances in Designed and Optimized Catalysts 12/31/2014
By Aminul Islam

Optimization of Cooling Systems 2/15/2015
By David Zietlow

Graphical Thermodynamics 3/1/2015
By Moufid Hilal

Momentum Press is looking for authors in this collection and in many more engineering and science areas. For more information on how to join the Momentum Press author team, please visit www.momentumpress.net/collections.

Announcing Digital Content Crafted by Librarians

Momentum Press offers digital content as authoritative treatments of advanced engineering topics, by leaders in their fields. Hosted on ebrary, MP provides practitioners, researchers, faculty and students in engineering, science and industry with innovative electronic content in sensors and controls engineering, advanced energy engineering, manufacturing, and materials science. **Momentum Press offers library-friendly terms:**

- perpetual access for a one-time fee
- no subscriptions or access fees required
- unlimited concurrent usage permitted
- downloadable PDFs provided
- free MARC records included
- free trials

The **Momentum Press** digital library is very affordable, with no obligation to buy in future years.

www.ingramcontent.com/pod-product-compliance
Lightning Source LLC
Chambersburg PA
CBHW070738220326
41598CB00024BA/3466